Dudley Grant Hays, Charles D Lowry

High School Laboratory Manual of Physics

Dudley Grant Hays, Charles D Lowry

High School Laboratory Manual of Physics

ISBN/EAN: 9783744763592

Printed in Europe, USA, Canada, Australia, Japan

Cover: Foto ©Paul-Georg Meister /pixelio.de

More available books at **www.hansebooks.com**

LABORATORY MANUAL

OF

PHYSICS

By

DUDLEY G. HAYS   CHARLES D. LOWRY   AUSTIN C. RISHEL

TEACHERS OF PHYSICS IN THE CHICAGO HIGH SCHOOLS

BOSTON, U.S.A.
GINN & COMPANY, PUBLISHERS
1895

COPYRIGHT, 1893
BY GINN & COMPANY

ALL RIGHTS RESERVED

PREFACE.

IN making this Manual two main objects have been kept in view. First, the teaching of Physics by the Inductive Method, that is, the presenting of a logically arranged course of experimental work that shall cover the ground of Elementary Physics. Second, the providing of sufficient laboratory work to meet the entrance requirements of any college in the country.

The authors are not so visionary as to suppose that boys and girls can, unaided, rediscover the laws of Physics, but we know that, if sufficiently careful directions are given to pupils in the performance of experiments, and definite instruction is given them as to the manner of *studying results* obtained, they will learn from Nature first-hand many of her great laws. And these will be much more strongly impressed than when learned from a text-book or from the teacher's experiments. From the generalizations made by the pupils, deductions can be made and then tested for their validity, thus keeping the pupil on the borderland between inductions and deductions, a place where the greatest mental development is obtained. This plan is the underlying idea of science study, and the acquiring of this *scientific method* is of far more importance than the mere gleaning of facts. The *facts* may in time be forgotten, but the method of mind operation will remain as part of the character developed.

Careful manipulation, accurate observation of phenomena, and logical deductions or generalizations should be the three steps kept in mind.

The notes written by the pupils should be *neat, terse*, and in *good* English, and should be logically arranged as shown in the steps mentioned above. Insist on these three elements of written work.

In the attainment of our second object, the requirements of Harvard, as being higher and more definite than those of any other institution, have been our standard, and we acknowledge great obligations to the Harvard Pamphlet.

Equivalents of nearly all their exercises will be found in this Manual.

With regard to the method of laboratory work, that of having duplicate apparatus so that a whole class may be performing the same experiment is of course the best. Much of the value of the work is in the discussion of the results in class-room. With this in view the apparatus selected is of the simplest possible kind that will secure good results, and most of it may be duplicated at small cost.

In the arrangement of topics, Magnetism has been put first because the experiments are easy, instructive, and fascinating, thus giving a very desirable introduction to the laboratory work. It also gives the teacher time to prepare his laboratory for the more difficult work which comes later. The other topics are arranged in about the usual order.

It is of course to be borne in mind that this is in no sense a text-book, nor intended to supplant one. It is simply a laboratory manual and may be used with any text.

LABORATORY MANUAL.

MAGNETISM.

The beginning work in the laboratory, covering the subject of Magnetism, can be performed by the pupils working in unison under the oral directions of the instructor, who can emphasize his directions in such a way as to leave no doubt as to the way to proceed. The pupils can thus be given a start in an earnest manner, and if the instructor is full of life, his pupils will soon gain a mental momentum that will count. Enthusiastic work coupled with neatness and accuracy should be the first aim. Unison work calls for but little apparatus here. Let each pupil bring two knitting needles, and the jeweller will gladly give you a pocket full of old watch springs which are excellent material for magnets. The magnetoscopes should be prepared by the instructor beforehand. Fasten a silk fibre to the bottom of a cork of a small bottle, such as a Florence flask, and to the lower end of the fibre fasten a short piece of magnetized watch spring. Clip the magnet with tinners' shears till it is balanced and is short enough to swing freely in the bottle. One for each couple of pupils will be enough. They can be kept throughout the year and will be of frequent use.

Rough notes can be taken in the laboratory and more elaborate writing can be prepared from them at home for the first few days. Nothing but *neatness* and *accuracy* of statements, based upon *facts observed*, should be accepted. A right beginning is half the battle.

EXERCISE 1. — Mix some tacks, some bits of brass, copper, soft iron wire, and some iron filings on a piece of paper. Stir the mixture with one end of a knitting needle, and note whether any of the objects cling to the needle. Now rub the needle from end to end several times, always in the same direction, on a piece of natural magnet or one end of a bar magnet — an electro-magnet arranged by the teacher can be used — and again stir the mixture.

Do any of the objects cling to the needle? What ones? Try to pick up a match; a pin; a sewing needle; a steel pen. What kinds of material cling to the needle? How does the needle differ now from its condition when first tried?

This attraction it manifests is due to the force called magnetism. The needle is now a magnet. What substances are attracted and held by a magnet when touched by it?

EXERCISE 2. — You have seen that when certain substances have been touched by a magnet they will cling to it. Ascertain whether it is necessary to touch these substances in order to show the attraction of a magnet for them. Cut out a piece of paper three inches long and two inches wide. Fold the ends together and crease the paper at the bend. Pin the two ends together with a soft iron nail parallel to the ends of the strip. Now pass a thread through the loop above the nail and hang it to any support, as, for example, over a ruler projecting from the edge of the table, so that the nail may swing freely. Hold a paper before one end of a magnet, and bring that end of the magnet near one end of the nail. Does the nail turn toward the magnet? Bring it near the other end of the nail, being careful to keep the paper between magnet and nail. Does the magnet attract each end of the nail even though it does not come in contact with it? How does this test of the effect of a magnet on soft iron or steel differ from that in Exercise 1? Is contact necessary for a magnet to attract soft iron?

EXERCISE 3. — Pour a row of iron filings on a piece of paper a little longer than the knitting needle magnet, so that you can lay the magnet on the paper and have it rest its entire length in filings. Roll the magnet back and forth in the filings and then pick it up carefully. Note whether the filings cling uniformly along the entire length of the magnet. Strip the filings off and repeat. Where do the filings cling? Try to pick up tacks at different points along the magnet. Stand a book on one end. Lay the magnet on the book so that one end — nearly half — of the magnet projects beyond the edge of the book. Now find how many tacks will hang in line at the end of the magnet. Try it for different places along the magnet, going from the end toward the center, the places being about a half inch apart. Record the results for each trial.

The places on a magnet where the filings cling in bunches, or where the greatest number of tacks cling, are called the poles of a magnet.

EXERCISE 4. — Get a heavy sewing needle — what is better, a short piece of watch spring one inch long — and dip it into filings to be sure it is not magnetized. Now lay it on a thin cork and float it in a dish of water. If the dish is small, it is well to oil the cork, which will then remain in the centre of the dish. When it has come to rest, with your pencil carefully turn it around so that it will be at right angles to the position it had when at rest, and note whether it will stay in about the position in which you put it. Now remove the needle and stroke it on one end of a magnet, always from end to end in the same direction. Then place it back on the cork and repeat the above work. What follows? Try other needles so as to be sure of the results. Does the same pole always seek the same direction? If one end always seeks the west, the pole at that end would be called the west-seeking pole. Give the correct names to the two poles of the magnets you have tried.

EXERCISE 5. — Cork a small clear glass bottle with a cork having a hole in it, through which passes a small glass tube, having a hook at the lower end to which is tied a silk fibre, on the lower end of which is fastened a piece of watch spring which has been magnetized, and which should hang horizontally in the bottle near the bottom. The tube can be moved up or down in the cork to adjust the height. The silk fibre allows the magnet to move freely. This little device is called a magnetoscope. Bring the point of a soft iron nail near the north-seeking pole of the magnetoscope. What follows? Do the same with the head of the nail. What follows? Now get a knitting needle magnet or a sewing needle magnet and try both ends of it the same way. What follows? How does the action when the nail was used differ from that when the magnetized needle was used? You thus see that the presence of magnetism in a body may be detected by use of the magnetoscope.

EXERCISE 6. — In the last exercise we saw that a magnetoscope could be used to detect the presence of magnetism. You saw that the effect of one end of the magnet on the north-seeking pole of the magnetoscope was the same as the effect of the point or the head of the nail, but that the other end of the magnet affected the north-

seeking pole of the magnetoscope in quite a different manner, — that, instead of the north-seeking pole of the magnetoscope being attracted in all cases, there was one case in which it was repelled. This *repellent* effect is the true guide to follow in determining whether or not a body is magnetized. Magnetize a piece of watch spring and float it on a cork and thus ascertain its north-seeking pole. Stick this end through a small bit of paper to label it. Proceed carefully and make the following tests, bringing one pole of the magnet slowly toward one pole of the magnetoscope, and record results :

Pole of magnet.	Pole of magnetoscope.	Effect.
North-seeking.	North-seeking.	?
North-seeking.	South-seeking.	?
South-seeking.	South-seeking.	?
South-seeking.	North-seeking.	?

By examining the above, discover a law of the effect of magnetic poles on each other. State it. How many poles has a magnet? What can you say in regard to these poles?

EXERCISE 7. — Magnetize a knitting needle, and by means of a magnetoscope determine its poles. Put a paper flag on the north-seeking pole. Break or cut the magnet in two at the middle, and roll the half which is labeled in iron filings. Does it seem to have two poles? Test this piece for polarity with the magnetoscope. What do you find? Now break this piece at the middle, and repeat the above tests. Results? You might continue this dividing process with the same results until you reached the smallest physical division possible which would give you a single molecule. What would be the condition of each of the molecules thus separated from a magnet? How do you think the molecules are arranged in a magnet, judging from what you have found in this experiment?

Lay one bar magnet on another so that like poles will be together. Have them project from some support above the table, — a book will do. Now bring a paper on which are tacks, or filings, up against the projecting ends. Note the quantity that clings to the magnets. Now superpose the magnets so that unlike poles will be together, and repeat the above. Results? Explain. If you had several flat magnets, how would you fasten them together so as to make one strong magnet?

EXERCISE 8. — Magnetize a short piece of watch spring or a sewing needle by rubbing it from one end to the other, always in the same way, with the north-seeking pole of the magnet, and then test the end of the newly made magnet which last left the old magnet for polarity. Try several and record results. Repeat by using the south-seeking pole of the magnet. Results? Can you discover a law for the arrangement of poles in magnetization? State it.

EXERCISE 9. — Make a feeble magnet of a piece of watch spring by stroking it once or twice with one end of a magnet. Test it for polarity by means of the magnetoscope. Place it on the table, and approach its north-seeking pole with the north-seeking pole of a bar magnet. Make several trials. Does the north-seeking pole of the small magnet jump towards the north-seeking pole of the bar magnet? If it does so several times, again test it for polarity. How do you account for this result? From this experiment do you think any magnet might have its poles reversed? Try several magnets made of watch springs, and see whether or not you can change the polarity at your pleasure. How can you weaken a strong magnet? Some magnets have their poles stamped or marked in the steel when they are purchased. Can we always rely on these marked names for the poles of a magnet?

EXERCISE 10. — Select a piece of soft iron wire or soft iron rod of small diameter (a soft iron nail will do) and test it with a magnetoscope to be sure it is not magnetized. Hold one end of the piece of soft iron in some iron filings. Place your index finger over the other end of the iron, and with your other hand bring one pole of a magnet down on the opposite side of the finger from the iron, the finger thus preventing the magnet from touching the soft iron. Keeping the objects in the position indicated, withdraw the soft iron from the filings. Do filings cling to the soft iron? Does it seem to be a magnet? Test it for polarity by means of a magnetoscope while the magnet is thus near it. How does the end of the soft iron farthest away from the magnet compare in polarity with the pole of the magnet used? Change ends of the magnet and try again. Change ends of the soft iron and repeat. If a piece of soft iron thus used manifests magnetism, it is called an induced magnet. What can you say about the arrangement of the

poles of an induced magnet in regard to the pole of the magnet used? If you should lay several soft iron nails down end to end in a line, leaving a small space between the ends, and were to hold the north-seeking pole of a strong magnet near one end of the line, what would all of the ends pointing towards the magnet be? All pointing away from the magnet would be what kind of poles?

EXERCISE 11. — Lay two rulers on the table about 4 in. apart and parallel to each other. Place a small magnet beween them, and cover the three objects with a piece of writing paper. Now with the thumb and finger sift some fine iron filings on to the paper from a height of about two feet. The outline of the magnet can be made more distinct, perhaps, by carefully tapping the paper with your lead pencil, being careful not to move the paper. The filings should show distinct lines radiating from the poles of the magnet. These lines of filings indicate the lines of magnetic force which constitute the magnetic field of the magnet. Make a careful sketch of this field.

Now proceed as before and map out the magnetic fields for a magnet; for the field when a north-seeking pole is opposite a north-seeking pole and about 1 in. from it; for the field when a north-seeking pole is opposite a south-seeking pole and about 1 in. away from it.

By using several horseshoe and several bar magnets so as to have the poles arranged around a central point, some interesting figures may be mapped out with filings.

MEASURING.

EXERCISE 12. — Measure the length and width of your table-top, and record the measurements in m. and decimals thereof, as,

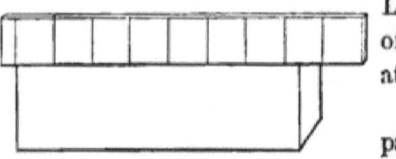
Fig. 1.

L. 2.145 m. Measure the height of your table, taking the average at several points.

Measure the length of this page, reading to quarters of mm. and record in cm., and fractions thereof, as, 27.95 cm. To measure accurately, the measuring rod must be so placed that the graduated edge will be in contact with the surface to be measured. It is also best to begin at some division a little from the end of the measuring rod, as indicated in Fig. 1.

EXERCISE 13. — Apparatus : Metric rule, exterior and interior calipers, sphere, rectangular prism and cylinder.

With the calipers get the diameter of the sphere, and by placing the calipers on the metric rule, get the diameter in cm., measuring to quarters of mm. Lest the ball be not a perfect sphere, take the average of several diameters. Find the vol. in ccm., using the formula on p. 150.

Measure the prism, and find its vol. in ccm.

Measure the cylinder, and find its vol. in ccm., using the formula on p. 150.

EXERCISE 14. — Place several thicknesses of paper in the micrometer calipers, and close the calipers upon them, using very gentle pressure. You will probably need the assistance of your teacher in reading your calipers. From the thickness of all, find the thickness of a single sheet.

Measure the thickness of the pieces of wire given you.

Measure the thickness of a bit of window glass, a sheet of mica, a hair, or any other object whose thickness you care to know.

EXERCISE 15. — Clean a small beaker, wipe it dry, place it on an accurate balance, and counterbalance it accurately by pouring sand into the other scale pan. Now draw from a burette exactly 25 ccm. of distilled water into the beaker, return the beaker to the balance, and find the weight of the water in g. What does 1 ccm. of water weigh? In case no burette is at hand, use a graduated vessel, but take a larger quantity of water, say 50 to 75 ccm.

PROPERTIES OF MATTER.

EXERCISE 16. — Select a two-hole rubber stopper which just fits a bottle. (If no rubber stoppers are at hand, an ordinary cork will do, if it be first wet, or, better, boiled in paraffine, to make it airtight.)

Through one of the holes pass the neck of a small funnel. Cover the other hole with a finger, and pour water into the funnel, pouring quite rapidly at first. Does it run into the bottle? Remove the finger an instant. What results? Why? What does this experiment show concerning water and air?

EXERCISE 17. — Fill a test tube half full of water, then, inclining the tube a little, slowly pour alcohol into it, letting it run down the side of the tube, being very careful not to shake the tube. When full, cover with the thumb and shake vigorously. Notice any change of volume.

Fill a test tube with water to within 2 cm. of the top. Now drop in about 2 or 3 ccm. of finely powdered sugar, a little at a time, and notice the volume when all the sugar is dissolved. Is the volume of the mixture equal to the sum of the volumes of the water and sugar? The same experiment may be tried with salt instead of sugar. What property of matter have these experiments illustrated?

EXERCISE 18. — Hold a piece of glass tubing, 30 cm. or more in length and 5 or 6 mm. in diameter, in the flame of a Bunsen burner so as to heat it about 10 cm. from one end. Hold the tube with both hands, and turn it steadily so as to heat it equally on all sides. Try to bend it from time to time to see if it softens. When thoroughly softened, remove from the flame and quickly stretch it out into a fine glass thread. When cool, break in the middle; or, if you pulled it apart, break off a few cm. of the finest end, and see if the thread-like part is still a tube. To do this, insert the small end in water and see if you can blow air through it. Examine it after withdrawing it from the water.

What property of glass allows you to draw it out thus?

EXERCISE 19. — Place a calling card on the top of a cork in a bottle. Upon the card, immediately over the cork, lay a penny. Now try to snap the card from under the penny so the penny will

drop upon the cork and remain there. A little patience will enable you to do it readily.

Explain why the penny remains behind.

Suspend a bag of sand weighing 2 to 4 lbs. by a piece of twine about 50 cm. long. Fasten a hook to the under end of the bag, and from this suspend about 50 cm. of the same kind of twine. Now take hold of the lower end of this second cord, and pull steadily downward till the cord breaks. Which piece of cord broke? Arrange as before, and this time break the cord by a sudden jerk downward. Which piece of cord broke? Why?

EXERCISE 20. — Into a flask which contains water to a depth of 3 or 4 cm. place a close-fitting cork containing a glass tube with its upper end drawn to a point, and the lower end reaching about 2 cm. into the water. The tube must be air tight in the cork.

Now blow into the flask until a considerable quantity of air has bubbled out of the lower end of the tube through the water. Remove the lips from the tube. What results? How did the compression to which you subjected the air affect its density? How did it affect the expansive force, or tension? Why does the water stop flowing?

EXERCISE 21. — *First Part.* Cut off a given length of solid rubber cord, or rubber tubing, say 40 or 50 cm. Now stretch it a little, release, and measure again very carefully. Repeat several times, stretching more each time, always measuring after stretching, to see if you can reach a point where the rubber does not return to its original length. A substance which uniformly resumes its original form is said to be perfectly elastic. Do you find rubber to be perfectly elastic?

Second Part. *For three pupils.* Fasten securely one end of a piece of spring brass wire No. 28, B. & S. gauge, 3 or 4 m. long to a screw or hook, or other support on the table-top, letting the wire extend over another table, if one table is too short. To the other end attach a 30-lb. Chatillon spring-balance.

Near each end of the wire fasten a drop of sealing-wax to serve as a marker. Lay a metric rule lengthwise under each end of the wire. Apply a tension of 1 lb. to the balance. This should stretch the wire straight. Now adjust each ruler under the wire so the side of the drop of wax is exactly over one of the divisions of the ruler, and do not allow the rulers to move throughout the experiment.

Increase the tension 2 lbs. (see Note 2). Read positions of both drops of wax to quarters of mm.; the difference between their movements is the amount the wire has been elongated. Reduce the force to 1 lb. and notice if the wire regains its original length.

Make several trials similar to the above, increasing the tension 2 lbs. at a time until permanent elongation is produced, remembering that after each trial you are to reduce the tension to 1 lb.

Tabulate your results as follows : —

Material of Wire, ... Gauge No., ... Length, ...

Force (F).	Movement of 1st Marker.	Movement of 2d Marker.	Elongation (E).
.... lbs.	.... mm.	.... mm.	.... mm.
.... "	.... "	.... "	.... "
.... "	.... "	.... "	.... "

Compare F and E, and see if there is any law connecting the stretching force and the elongation.

Note 1. — Very much of your succeeding work will be to derive such laws. In doing so, there are two general ways of comparing the results of experiments : (1) If the *quotients* (ratios) of corresponding results are always the same, then the results are said to be *directly proportional* to each other; for example, if in your experiment $E \div F$ (or $F \div E$), always gives the same quotient, then we say that the stretching force is *directly proportional* to the elongation. (2) If the *products* of corresponding results are always the same, then the results are *inversely proportional* to each other; that is, if $F \times E$ gives always the same product, we shall say the stretching force is *inversely proportional* to the elongation. In most cases the quotients or products will not be exactly equal, owing to errors in work. If the differences make only a small part of the whole results, the law may be stated. If the variations are considerable, state what you think the law to be, and say it is approximately proved. (3) Sometimes one set of results is to be compared in either of the above ways with the squares or square roots of another set of results, and occasionally with the cubes or cube roots.

Note 2. — The spring-balances are made to be used with the hook hanging vertically downward ; in any other position the index stands back of the zero mark. In this case, a "balance correction" must be added to your reading. To find the balance correction for any position, hook a more delicate balance to the draw-bar, and see what force is necessary to bring the index to the zero mark ; or, a cord attached to the hook may be passed over a smooth-running pulley. To the cord attach a very light scale-pan on which to lay weights sufficient to move the index to the zero point. The weight of the scale-pan and weights is the correction sought. In all cases where accuracy is sought, the balance correction should be made.

EXERCISE 22. — Support two of the prisms described on p. 36 upon blocks 3 or 4 cm. high, so their upper edges shall be 1 m. apart; upon these lay a piece of straight-grained pine or basswood 3 ft. 6 in. long, $\frac{1}{2}$ in. thick, and 1 in. wide. Alongside the middle of the rod support another of the prisms at the same height, but parallel to the rod and 5 cm. from it. Upon this prism, as a fulcrum, rest a very light rod 32 cm. long, so its end projects 2 cm. under the first rod. The arrangement of the apparatus should be as shown in Fig. 2.

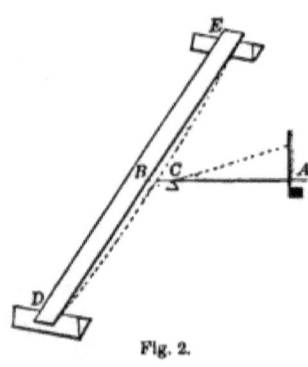

Fig. 2.

Press downward upon the middle of the rod DE. A will rise; and since AC is five times the length of CB, A will rise five times as far as B is depressed, and the amount of rise at A will serve to show the depression at B. Consider B as being the point where the rod DE touches the pointer AB.

Measure the height of A above the table, reading to halves of mm. Now lay a weight of 200 g. upon DE at B, and again get the height of A. Remove the weight and measure again. Now use weights of 400, 600, 800, and 1000 g., reading the height of the pointer in every case, as before, and stopping at any time permanent bending is produced. Great care should be exercised in handling the weights so the apparatus is not jarred and the position of the pointer changed.

In case other than flat weights are used, they may be laid on a pan suspended from the under side of DE by a string which passes through a hole in the table-top.

Tabulate the results as follows : —

First case. — Rod flat; supports 100 cm. apart.

LOAD.	READINGS WITHOUT LOAD.	READINGS WITH LOAD.	RISE OF POINTER IN mm.	DEPRESSION OF ROD IN mm.	DEPRESSION PER 100 g. IN mm.
.... g.	....	....	....	....	....
.... "	....	....	....	....	....
.... "	....	....	....	....	....

Second test. — Support the same rod on its narrow side with supports 100 cm. apart, and load from 500 to 2000 g., increasing 500 at a time, reading as before.

Third test. — Use a rod of same material and grain, but 1 cm. thick and 1½ cm. wide, laid flat, with the supports 100 cm. apart. Use weights of 100 to 500 g., increasing 100 g. at a time.

Fourth test. — Use the same rod laid flat with supports 50 cm. apart, with forces of 200 to 1,000 g., increasing 200 g. at a time.

A study of your tabulated results should enable you to answer these questions: —

1. What is the relation between the force employed and the amount of bending of an elastic rod? Refer to Note 1, p. 20.

2. What is the relation between the length of the rod and the amount of flexure, *i. e.*, depression of its center?

EXERCISE 23. — Arrange apparatus as shown in Fig. 3. A is a circular board 12 in. in diameter and 1 in. thick. B is a rod of ash or hickory, ½ in. × ½ in. in cross section, and 42 in. long, clamped with its center exactly over the center of A. The support of C should be adjustable to heights from 50 to 100 cm. above A. The rod is firmly clamped to C. It is best to have the point of a wire nail project downward from the center of A into a hole bored to receive it in the table-top, so as to prevent A from moving about. D is a cardboard with an arc of 60° graduated to single degrees; the radius should be 6½ in. or 7 in., and its center should coincide with the center of A.

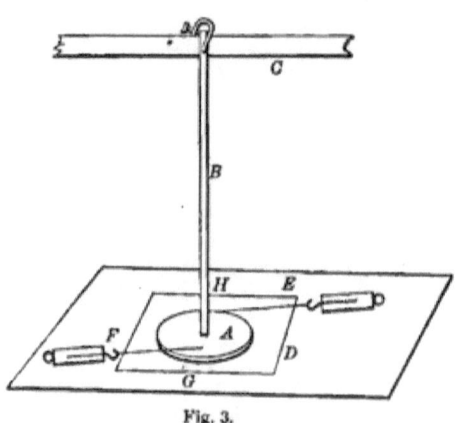

Fig. 3.

1. Adjust C so there shall be a space of 100 cm. between the clamped portions of the rod, and apply force with two 4-lb. Chatillon balances attached to the cords E and F, pulling in opposite directions, with equal force, and keeping the cords always at right angles to the line GH. Use forces of 2, 4, 6 and 8 oz. and record for each force the number of degrees of torsion, as read on the card D.

2. Now lower C so that but 50 cm. of B are exposed to torsion, and use forces of 4, 8, 16 and 32 oz.

3. Substitute a similar rod of $\frac{3}{4}$ in. $\times$ $\frac{3}{4}$ in. sectional area, and with a length of 100 cm. use forces of 1 lb. to 4 lbs.

4. Use but 50 cm. of the large rod, and forces of 2 lbs. to 8 lbs., using larger balances.

If in any cases the forces suggested are not suited to the strength of the rods, use forces that are suitable.

Tabulate the results, and from a comparison of results answer the following questions: —

1. What is the relation between the force employed and the amount of torsion?

2. What is the relation between the length of the rod and the amount of torsion?

3. How nearly do your results agree with the law, "Torsion varies inversely as the fourth power of the diameter of the rod"?

EXERCISE 24. — *Apparatus:* about 1 m. of spring brass wire, No. 28 B. & S. gauge, and a 30-lb. Chatillon spring-balance. The teacher should prepare the balance for use by slipping an empty thread spool over the hook so that when in position and the balances are stretched, the axis of the spool is at right angles to the draw-bar of the balance. Choose a spool of such length that when properly cut away at the ends it cannot turn on the hook.

One end of the wire should be wrapped several times around a cylindrical support, not less than 2 cm. in diameter. A gas-pipe, water-pipe or table leg will answer. After wrapping the wire about the cylinder, fasten the end to a tack or screw driven near the cylinder.

The other end of the wire is to be fastened to the eye in the end of the draw-bar of the balances, after which the wire should be wrapped several times around the spool on the hook of the balance, care being taken all the time not to twist or kink the wire. Be careful to keep your hands and face out of danger in case the wire recoils after breaking.

The apparatus being ready, the purpose is to determine the amount of force necessary to break the wire. Begin with a small strain on the wire, holding the eyes over the balance so that the index reading of the balance may be accurately taken. Increase the force steadily, noting all the time the reading of the balances,

until the wire breaks. Repeat three or four times, using a different piece of wire each time, and thus get the average breaking strength of the wire, making the proper balance correction as noted in Exercise 21.

Now test a wire of the same thickness but of different material, as iron, copper, or steel.

Returning to the first case, weigh a given length of the brass wire, as 1 or 2 m., and determine what length of it would just break of its own weight if suspended by one end.

EXERCISE 25. — Cut a lead bullet or small piece of sheet-lead in two with a sharp knife, then press the freshly cut surfaces together with a twisting motion until they cohere. What do the facts that a smooth surface and pressure are necessary indicate about the distance through which cohesion acts?

In same way unite freshly cut surfaces of rubber, paraffine or wax.

Place a small drop of mercury on the table, and note carefully its shape. Why does not the mercury spread out over the table? Would the great weight of mercury hinder or assist it in keeping a spherical form?

Place a drop of water on a clean pine board, and account for the shape of the drop. Now place drops of water on a board that is covered with powdered rosin. Note and account for the difference between the shape of these drops and those of water on a clean surface. What force holds the powder to the surface of the drop?

EXERCISE 26. — Suspend a disc of tin or sheet-zinc 5 to 8 cm. in diameter by means of three threads from the hook on the under side of the scale-pan of a specific gravity balance. Now support the balance at such a height that when the beam is horizontal, the disc rests lightly upon the surface of the water in a battery jar or tumbler. Be sure there are no air bubbles under the disc. Carefully lay weights on the other scale-pan, — do not *drop* them in — and find the force necessary to tear the disc away from the water. Look now at the under side of the disc and decide which is the greater, the adhesion of the water to the disc, or the cohesion of the water. Was the force overcome that of cohesion or of adhesion?

Wipe the disc dry, coat its under surface with oil, and again find the force necessary to tear it from the water.

Exercise 27. — (*To the teacher:* Sets of capillary tubes, cleaned for use, should be in readiness; tubes 1, 2 and 5 mm. in diameter and 20 cm. long are suitable. Liquids for cleaning them should be in readiness so the pupils lose no time. A rubber bulb with rubber tube attached will assist in drawing the liquids through the tubes. To clean the tubes, wash first in dilute nitric acid, rinse in water, then wash in dilute caustic potash or caustic soda, then rinse in water.)

In a tumbler or beaker place distilled water about 6 or 8 cm. deep. Lower the smallest tube to the bottom of the tumbler, so as to wet the tube inside. First study the form the water takes around the tube, and around the sides of the tumbler. What form of attraction is displayed here? Now lift the tube entirely out of the water, dry the end, lower it very slowly toward the water and watch to see if the water jumps to meet the tube. Incline the tube a little, so one corner approaches the water first.

Now lower the tube about 2 cm. into the water, holding it vertically, and notice at what height the water stands in the tube. Lower still more, and see if the water in the tube remains at the same height above the water in the tumbler. Now lift the tube entirely out of the water, and notice if any water remains in the tube, and if so, to what height. Arrange a table similar to the following, and with the other tubes and liquids determine the facts: —

Diameter of Tube.	Height in Cold Water.	Height in Hot Water.	Height in Alcohol.
1 mm.			
2 mm.			
5 mm.			

When the large tube is used, what is the shape of the upper surface of the water in the tube?

Pour clean mercury into a test tube to the depth of about 5 cm., dry the tubes and place them one at a time into the mercury, holding the tubes at the side of the test tube so you can see if the mercury enters the tubes. Record your observations. Make a sketch showing the shape of the mercury surface within and around the largest tube.

Take two plates of glass about 3 × 8 cm. Slips for mounting microscopic specimens are convenient. Lay the one on top of the other with a bit of wood 2 mm. thick separating them their entire length along one edge. Slip a rubber band around them so as to hold them firmly together, and lower them vertically into the water. Sketch them, showing the height to which the water rises between the plates.

These phenomena are said to be due to capillary attraction. Can you explain them by means of forces already studied?

PENDULUMS.

Exercise 28. — Suitable pendulums are easily made from lead bullets by cutting into one side with a knife, inserting a thread and pinching the lead tight about the thread with a pair of forceps. Iron balls with holes through them will also serve. These pendulum bobs should be in readiness for the class. The exercise can be worked by a class in unison if desired.

Measure a pendulum 60 cm. long. In taking its length, measure from the point of support to the centre of the bob. The pendulum may be suspended between the thumb and finger, if you are careful to give it a definite support so it does not change its length during the experiment. It may also be held in a pair of forceps, hung over the back of a knife-blade, or suspended in a variety of ways.

Start the pendulum vibrating through an arc of not over 10 or 15 cm., and count the number of *single* vibrations in a minute. Repeat two or three times before recording. Now, keeping the pendulum still the same length, let it swing through a much larger arc, say 30 to 40 cm. Repeat and record. What influence has the length of the arc upon the number of vibrations made in a given time?

Substitute a cork bob, or other light body for the metal bob, retain same length as before, and determine whether the material of which the pendulum is made affects its rate of vibration.

Arrange a table as below, and, using the metal bob, record your results after two or three trials with pendulums of each length :

No. Vibr. per Min.	Period, i. e. Time of 1 Vibr.	Length.
....	....	25 cm.
....	....	100 "
....	....	50 "
....	....	200 "

By studying your results determine the law connecting the period of a pendulum with its length. (Note 1, p. 20.)

MECHANICS.

For applying force in mechanics, the Chatillon 4-lb. spring-balance, graduated to ounces, is recommended. Where weights have to be provided, bags of sand are both convenient and cheap. The bag should be made of closely woven material, such as ducking.

Suitable levers may be made from lattice sticks 13 in. long, 1 in. wide and $\frac{1}{4}$ in. thick. Half an inch from each end a line should be drawn across the broad side, and between these, 11 other lines 1 in. apart; the stick now resembles a foot rule. On the other broad side, exactly opposite each line, a small notch should be cut across the lever. A triangular prism 1 in. long and $\frac{3}{4}$ in. high, when lying on one face, will serve as a fulcrum. Where metric weights are not at hand in sufficient numbers, nickels will serve as 5 g. weights, and 10 g., 15 g., and 20 g. can be cut from sheet-lead.

Sets of two single, and two double pulleys should be provided; also stout linen or cotton cord for same. Fishline is good.

It is recommended that Exercise 40 be performed before the class, they making their own observations and deriving the laws unaided so far as possible. A wire 2 or 3 mm. in diameter should be stretched across the room on a grade of 1 to 16; by means of a turnbuckle it can be stretched very tight. On the wire is a smooth-running pulley carrying an iron weight of 2 lbs. so arranged as to roll down the wire when released. The weight is held in place at the upper end of the wire by means of an electromagnet. If the friction prevents the weight from moving 1 foot the first second, and corresponding integral numbers of feet for each succeeding second, the inclination of the wire should be increased till these results are reached. A seconds pendulum, hung by a copper wire, is so arranged that a very thin wire pointer on the under side of the bob cuts through a ridge of mercury (a trough 3 mm. wide heaped full) at the middle of its arc, thereby closing a circuit which rings a bell or clicks a telegraph instrument once a second. The ringing of the bell breaks the circuit of the electromagnet, thus releasing the weight. Above the wire, and parallel to it is stretched a cord to which are pinned bits of paper 1 foot apart, to mark the distance the weight moves.

EXERCISE 29. — For three pupils. On a smooth, square board rule off a square 12 in. on each side. Divide this square into 36 two-inch squares, and at the intersections of all the lines bore $\frac{1}{8}$ in. holes not quite through the board. Number these holes from 1 to 49. Provide 3 iron pegs fitting the holes pretty closely. (Wire nails broken off will answer.)

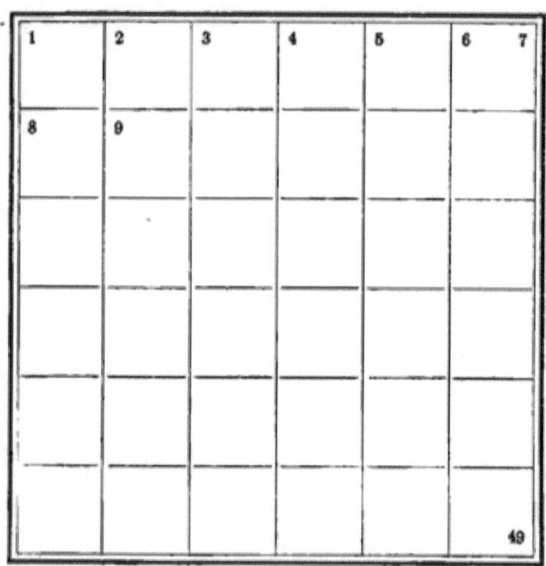

FIG. 4.

Lay three smooth marbles of equal size upon the table, and upon these the board, marked side up. Attach a cord a foot long to each of the pegs, and insert the pegs in three holes in any one line. Hook a spring-balance into the end of each cord, and apply forces at *right angles* to the line on which the pegs stand, so that when all the balances are stretched, the board will not move. Be sure the cord does not bear down upon the board, and that it lies exactly over the line traced on the board. Hold the balance firmly in both hands, with the knuckles resting upon the table, and keep the head in such a position that at a given signal all can read the balances instantly without moving anything but the eyes. Remember that the percentage of error is much less if the force applied is considerable than if it is small.

Now, having produced equilibrium and taken the readings, make a diagram of the board in your note-book, number the holes, and record the conditions which produced equilibrium. The record can be briefly stated in the form of a table, thus: —

Hole Number.	Force.	Direction: N., S., E., or W.
1. {	....	....

Now change the distance between the pegs, and record the conditions which produce equilibrium. Be sure that in some of the trials the pegs are at unequal distances from each other.

Return now to the original case of equilibrium; let all the forces retain the original direction and magnitude, and two of them the original points of application. Seek a new point of application for the third force which shall leave the system in equilibrium as before. Take each of the other two forces as the roving one, and record the results of your trial, together with any conclusions you may have reached.

Again, returning to any of the cases of equilibrium, see if the equilibrium continues while the forces, keeping the same magnitude and points of application, and still remaining parallel to each other, are turned about into new directions, *i. e.*, are no longer at right angles to the line of pegs.

Now, by referring to your results, answer the following questions: —

1. What is the direction of the equilibrant of two parallel forces acting in the same direction?

2. What is the magnitude of this equilibrant?

3. What relation exists between the magnitude of the two forces and the distances of their points of application from the point of application of the equilibrant?

EXERCISE 30. — Lay the prism on the table, and on its edge place the middle notch of the lever. This support on which the lever turns is called the *fulcrum*. If the lever does not balance, lay a lead chip on it at such point as to produce equilibrium. Note

that this lever balances when neither end shows a tendency to rise after being depressed.

On the right-hand end of the lever place 5 g. 6 in. from fulcrum; *in all cases* place center of weight exactly over the *line* on the *lever*. Call this 5 g. the *weight* to be lifted, and find what *power* must be applied six in. to left of fulcrum to produce equilibrium. Designate the weight and power by W and P, the fulcrum by f. The distance from W to f is called the *weight distance*, Wd; that from P to f, the *power distance*, Pd. Place your results in the proper place in the following table, and by use of the lever find all the missing terms in the table. For convenience, keep Wd to your right: —

	P.	W.	Wd.	Pd.
1	?	5 g.	6 in.	6 in.
2	5	10 "	3 "	? "
3	?	15 "	4 "	3 "
4	10	? "	2 "	4 "
5	10	15 "	? "	6 "

Now place the fulcrum under the notch 4 in. from one end, and balance the lever with lead chips; then find the missing terms in the following table: —

	P.	W.	Wd.	Pd.
1	5	10	4	?
2	10	?	3	6
3	?	20	4	8

Study these tables to see if you can find a general relation between P, W, Pd, and Wd.

Moments of Force. The moment of a force is the tendency of that force to produce rotation about a fixed axis, and is computed by taking the product of the *force* times the *distance* of its point of application from the axis of rotation. Since P and W tend to produce rotation about the fulcrum in opposite directions, call that distance *positive* whose moment tends to motion in a direction *with* the hands of a watch, and the other *negative*. Now, compute the moments in all the preceding cases, arrange the results as

indicated below, being careful of algebraic signs, and find the algebraic sum of the moments in each case: —

1. $\left\{\begin{array}{l}\{-\ Pd.\ \}\ \{\ P.\ \} = -\ ? \\ \{+\ Wd.\ \}\ \{\ W.\ \} = +\ ?\end{array}\right\}$ Sum of moments. ?

2. Etc.

Now calculate the moments in the following case: With f 4 in. from end, W 20 g., Wd 4 in., balance by applying two weights of 5 g. and 10 g. at different places on the other arm.

EXERCISE 31. — Arrange two levers as shown in the figure,

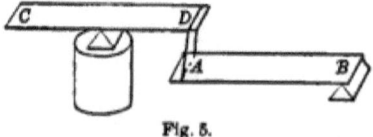

Fig. 5.

using a loop of thread to connect them. Balance the lever with lead chips.

AB is the lever under consideration. Since you learned in the preceding exercise that a power applied downward at C exerts an equal upward force at D or A, you may in each of the succeeding tests treat the power as applied upward at A, although you apply it over C. On AB place a weight of 20 g. 6 in. from B. What is the Wd? Pd? Find what power applied at C will produce equilibrium. Fill out the following table:—

	P.	W.	Pd.	Wd.
1	5	?	12	6
2	?	15	12	4
3	?	{10, 15}	12	{6, 4}
4	10	20	10	?

EXERCISE 32. — Arrange the levers as shown in the figure. A pin may be driven through the half-inch mark at A, bent into a

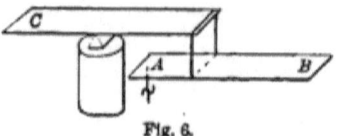

Fig. 6.

hook, which is then hooked into another bent pin driven into the table-top. Balance the levers. AB is the lever under consideration. Where is its fulcrum? If power be applied at C, where does it exert itself upon AB? Lay a weight

of 5 g. on the half-inch mark at B, and apply power at C to produce equilibrium. What is the length of the power arm on the lever AB? The weight arm? Arrange your results in the first line of the table. Find the other missing terms:—

	P.	W.	PD.	WD.
1	?	5	?	?
2	20	10	4	?
3	?	{5, 5}	4	{6, 10}
4	20	?	3	10

Compute the moments.

EXERCISE 33. — Take a wheel and axle device; measure the diameter of both the wheel and the axle; call them D and d. What is the ratio of D to d? What is the ratio of their circumferences?

Hang a weight of 3 to 5 lbs. on the cord passing around the axle, hook the spring-balance into the cord passing around the wheel, and raise the weight, pulling vertically downward. Read the balances to quarter ounces while the weight rises, and add the balance correction for this position (page 20, Note 2). Now let the weight slowly descend, reading and making correction as before. The average of these two readings is the power necessary to support the weight on the axle. Repeat your readings several times before recording. Call the power and weight P and W. What is the ratio of P to W? Compare this ratio with that of the diameters. Can you see a general relation between P, W, D, and d?

How far must the power act to raise weight 10 cm.? Do you find it necessary to measure to answer this question?

In this exercise, which is gained, intensity or velocity? Which was lost?

Apply W to wheel and P to axle, and state which is gained and which lost, intensity or velocity. Calculate the work done by P in raising W 4 in.* Compare this with the work necessary to raise W directly upward the same distance.

EXERCISE 34. — Arrange your apparatus as in Fig. 7 (a), using a weight of 3 to 5 lbs. Hook a spring-balance which registers to

---

* The work done by any power is the product of the power times the distance it has moved.

ounces into the cord, and raise the weight slowly, applying the power as nearly vertically downward as you can; note on the balances the power used, reading to half ounces. To this reading add the balance correction for this position of the balance. Now allow the weight to descend slowly and note the power used as before. Repeat several times; then take the average of the power used in raising and lowering as the force necessary to support the weight.

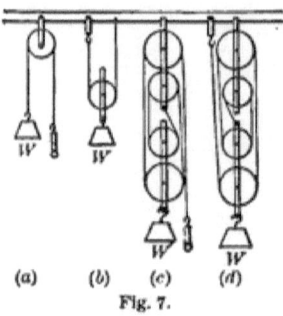
Fig. 7.

1. What is the ratio of P to W?
2. How far must P act to raise W 6 in.?
3. What is the ratio of Pd to Wd?
4. In this arrangement, how many sections of the cord support the weight?

Arrange the pulleys as shown in the other figures, and answer the above questions for each arrangement. When done, arrange the results of the four trials in the following table:

	P.	W.	Pd.	Wd.	SECTIONS OF CORD.
1	....	....	....	....	....
2	....	....	....	....	....
3	....	....	....	....	....
4	....	....	....	....	....

Can you, by observing P, W and the number of sections of the cord, discover a general law for the pulley?

Compute the work done by the power in the last case in raising the weight 10 in., and compare this with the work necessary to raise the weight 10 in. directly upward.

EXERCISE 35. — Support a smooth board about 80 cm. long and 30 cm. wide firmly on the table with one end raised from half to two-thirds the length of the board above the table. The board so arranged is called an inclined plane.

Measure the length of the plane. Measure its height. This is the vertical distance from the under edge of the upper end of

the board to the table. Call these L and H. Find the ratio of H to L.

Place a weight of 2 to 5 lbs. on the cart,— a roller skate will do,— and weigh the cart and its load. Call this weight W. Place the loaded cart on the plane, and apply power with the spring-balances so as to draw it steadily up the plane, being careful to apply the power in a direction parallel to the plane. Take very careful readings— at least to half ounces — as the weight ascends, making the proper balance correction. Now allow the cart to roll down the plane, taking readings at it moves. Repeat several times; then take the average of the readings while ascending and descending as the power necessary to support the cart on the plane, irrespective of friction. Call this P, and find the ratio between P and W. Compare this with ratio of H to L.

How far must the power act to draw W the full length of the plane? How far has W risen vertically in the meantime? Calculate the work done by P in raising W 8 in., and compare with the work necessary to raise W 8 in. directly upward.

EXERCISE 36.— Place a smooth pine board about 1 m. long upon the table. Take a smooth pine block about $2 \times 6 \times 15$ in. At the center of one end of the block fasten a screw-eye with cord attached.

Place the block with one of its narrow faces upon the board, and with the balances move it slowly and steadily along the board, with its fibers parallel to those of the board, being careful to keep the cord parallel to the table-top. Record the force necessary to *start* the block; now find the force necessary to keep it in motion. As the force necessary to move the block may vary at different parts of the board, a place should be sought where for several inches the force is uniform, and all readings taken there. Repeat the process several times, take careful readings, make the balance correction for this position and call the average of these corrected readings the force necessary to overcome friction. What is the amount of pressure of the block upon the board? The ratio of friction to pressure, *i. e.*, friction divided by pressure, is called the *coefficient* of friction. Find the coefficient in the above case.

Lay the board on its broad side, and find the coefficient. Now place a weight of 2 or 3 lbs. upon the block, and find the coefficient.

Lay the block upon two round lead-pencils and find the coefficient of rolling friction.

EXERCISE 37. — Through one corner of a piece of cardboard, 6 or 8 in. across, prick a hole with a pin and enlarge the hole so the card will revolve freely about the pin. Hang a plumb-line from the pin, and suspend both the card and plumb-line. It will be well to drive the pin horizontally into a wooden support. When both have come to rest, place a mark very carefully back of the plumb-line just where the line leaves the card. Remove the card, and with ruler and sharp pencil trace on the card the line of direction of the pendulum.

Suspend as before from any other position and trace the line as before. At the intersection of these two lines stick a pin through the card, bend it into a hook and suspend the hook by a thread. If the card hangs horizontally, the center of gravity lies half way between the two surfaces at the intersection of these lines.

Remove the hook, find a new line of direction and see if it intersects the two other lines at the point where the hook was.

EXERCISE 38. — *Influence of Weight of Lever.* *Apparatus:* Lever and fulcrum used in Exercise 30; metric weights and ruler.

Find the center of gravity of the lever by balancing it over a knife edge, or other sharp support. Weigh the lever to tenths of a g. Place the notched face of the lever upward, lay a weight of 20 g. on the half-inch mark at one end, and carefully adjust the fulcrum so the lever will balance without using additional weights. Be sure the fulcrum is *exactly cross-wise* under the lever. Now measure very carefully the distance from weight to fulcrum, also from center of gravity of lever to fulcrum. Compute the moment on the weight side. Find, by applying the law of moments, at what distance from the fulcrum the weight of the lever, if applied at a single point, would have to act to produce the equilibrium observed.

Now place the weight on the second mark, balance, measure, and compute as before.

Again, increase the weight to 30 or 40 g., and proceed as before.

Consider now whether in every case the weight of the lever has acted as if applied at the same point; and, if so, what is that point?

EXERCISE 39. — On a drawing board fastened firmly against a wall, suspend two spring-balances, reading to ounces, as shown in Fig. 8.

On the cord *BD* hang a weight of 2 to 5 lbs. With thumb-tacks or pins fasten a sheet of paper to the board, with the middle of the paper back of the intersection of the cords; using a ruler and sharp lead-pencil, trace on the paper lines parallel to each of the three cords, being sure the lines intersect at a common point.

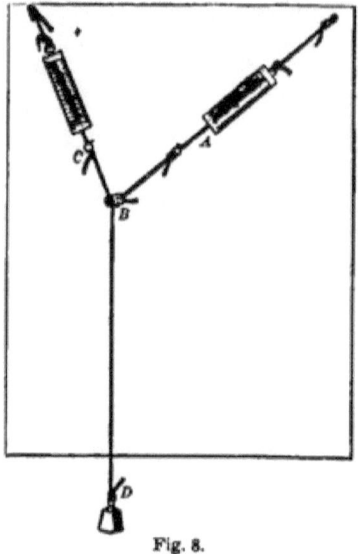

Fig. 8.

Read the balances, and beside each line mark the reading in ounces. Now remove the paper from the board, decide on some scale suited to the force used and size of the paper, and mark off on each of the three lines a distance corresponding to the force applied in that direction. Consider *AB* and *BC* as sides of a parallelogram; finish the parallelogram and draw its diagonal *BE*. On the scale adopted, how great a force does the diagonal represent? Compare this with *BD*. Considering *AB* and *BC* as components, what are *BE* and *BD*? What truth does this experiment seem to teach?

EXERCISE 40.* — Let the weight roll down the wire, and notice how far it moves in one second. Repeat several times. Then find the distance the weight moves in two seconds, etc. Tabulate the results:

Length of plane .... ft. Height of plane .... ft.

Entire distance passed over in
One second .... ft.
Two seconds .... "
Three   "   .... "
Four   "   .... "
Five   "   .... "

By a comparison of the above results, fill out the following table: —

Distance passed over in
First second .... ft.
Second "   .... "
Third   "   .... "
Fourth "   .... "
Fifth   "   .... "

* For directions as to apparatus, see p. 36.

Let the distance passed over during the first second be represented by d. Now at the right of both the above tables, write the number of d's passed over in each of the periods of time given in the tables.

Compare the number of d's passed over in any number of seconds with that number of seconds, and state a law, if you can. (Law I.)

In the second table, compare the number of d's passed over during each second with the number of that second, and state a law, if you can. (Law II.)

The velocity at the beginning of the first second was 0; it increased uniformly throughout the second; the average velocity for the second, therefore, equals half the sum of the velocities at the beginning and end of the second. But the space passed over during the first second equals the average velocity for that second. How many d's represent the velocity at the close of the first second?

Now, observing that the velocity at the close of the first second is also the velocity at the beginning of the second second, and also that the number of d's passed over during the second second is the average velocity for the second, find the velocity at the end of the second second. Proceeding in the same way, make a table for the velocity at the end of each second.

The velocity gained each second is called the acceleration or increment of velocity for that second. How many d's represent the increment for each second? State a law for the velocity at the end of any second. (Law III.)

Let S represent the distance passed over in any number of seconds by a body free to move, acted on by a constant force; s, the distance passed over in any single second; t, the time, and d, as before, the distance passed over during the first second. State Laws I, II, III by means of formulae; e. g., $S = ?$ $s = ?$ $V = ?$

The force of gravity is the constant force here employed, and these laws will apply to freely falling bodies, provided you multiply your value of d by the ratio of the length of the incline to its height.

HYDROSTATICS.

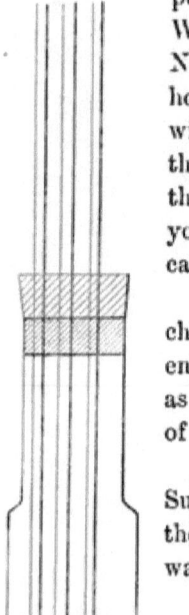

Fig. 9.

EXERCISE 41. — Put a wet card on one end of a chimney and push it down into a glass jar filled with water. What keeps the card from leaving the chimney? Now carefully pour water into the chimney and note how the height of the water in the chimney compares with the level of the water outside the chimney when the card is pushed off. Make several trials, and vary the depth to which you push the chimney. From your results, to what is the force which holds the card against the chimney equal?

Tie a piece of rubber tissue over one end of a chimney and after filling it with water cork the other end with a cork through which pass three glass tubes as shown in Fig. 9. Be careful to have the centers of the lower ends of the tubes on the same level.

Note the heigth of the water in the three tubes. Subject the liquid to pressure by pressing gently on the rubber. Note the relative heights to which the water is pushed.

How did the pressure get to the water in the tubes?

In what directions did the water move when entering the tubes?

What facts does this exercise show in regard to the transmission of pressure to which water is subjected?

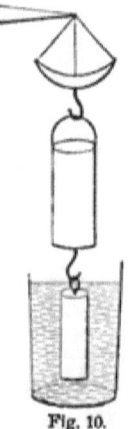

Fig. 10.

EXERCISE 42. — Use a small brass bucket and plug. You will notice that the plug exactly fills the cavity of the bucket. Now hook the plug to the bottom of the bucket and suspend the bucket from one scale-pan of a balance, then put weights, or shot, or sand into the other scale-pan of the balance till equilibrium is restored. Now place a dish of water under the plug and elevate the dish (or lower the upright standard of the balance) till the plug is entirely submerged *when the arms of the balance are held horizontal.* Release your hold. Is the equilibrium disturbed? Carefully pour water into the bucket till the arms of the balance are horizontal. The apparent

loss of weight of the plug when it is immersed in the water is due to the buoyant force of the water which it displaced. How does the volume of the water displaced by the plug compare with the volume poured into the bucket? Judging from this, the buoyant force of a fluid upon a body submerged in it is equal to the weight of what?

Did the plug lose any of its weight? Did it lose weight so far as the balance reading was concerned? It is customary to speak of a body as losing weight when it is immersed in a liquid, but this is a peculiar use of the word "lose."

The truth you have learned in this exercise is called Archimedes' principle.

SPECIFIC GRAVITY.

EXERCISE 43. — Measure very carefully a rectangular block about 3 cm. × 5 cm. × 10 cm. Compute its volume. Weigh it. What is the weight, in g., of 1 ccm. of this wood? This number is called the density of the wood. Balance, with sand, a beaker, then pour into it from a graduate as many ccm. of water as will equal the volume of the wood. Weigh. Divide the weight of the wood by that of the water. The quotient is called the specific gravity of the wood. Formulate a definition for specific gravity.

EXERCISE 44. — Weigh an irregular piece of some solid denser than water and not soluble in it. Tie a fine thread to the solid, attach it to one side of the balance and let it be immersed in water. Re-weigh. By Archimedes' principle determine the volume of the solid. What is the weight of the same volume of water? What, then, is the specific gravity of the solid?

In like manner find the specific gravity of at least 6 substances.

Tabulate as below where W equals weight in air; W' equals weight in water; D equals difference between your result and that given in the table of specific gravity in the back of this book. This difference is not necessarily an error, as the result given in the table is the average of many specimens:

NAME.	W.	W'.	LOSS.	SP. G.	D.

EXERCISE 45. — Weigh a solid less dense than water, e. g., a piece of paraffine, wood or cork that has been coated with oil or hot paraffine to make it water-proof. Attach to it a sinker and weigh them together in water. Weigh sinker in water. From these three weights find the specific gravity of the body tested.

EXERCISE 46. — Find the specific gravity of a liquid by means of a specific gravity bottle. The amount of water the bottle will hold is usually cut in the glass, and need not be tested by the pupils. If the bottle has no counterpoise, balance it, when perfectly dry, with sand. Fill the bottle with the liquid to be tested,

weigh, and calculate the specific gravity. In filling, be sure no air bubble is under the stopper; wipe the outside dry, and be very careful that none of the liquid is brought in contact with the scale-pan. If a second liquid is to be tested, rinse out the bottle with a little of this liquid, then proceed as before.

EXERCISE 47. — Obtain a narrow U tube having arms about 50 cm. long. Pour into it water until it stands about 20 cm. high in each arm. Next pour into one arm some liquid that will not mix with water, as kerosene, mercury, etc. Measure the vertical distances from the surface where the liquids meet to the free surfaces of the liquids. In this, as other experiments, measure to the level of the upper surface, not to the upper edge of the meniscus.

From your measurements, calculate the specific gravity of the liquid tested.

Alternative: — Connect, by means of rubber tubes, to each arm of a Y tube, a glass tube about 50 cm. long. Over the stem of the Y slip a rubber tube about 20 cm. long. Unless the rubber fits the glass closely, wrap with soft cord. Insert one of the tubes into a beaker of water; the other, into a beaker containing some other liquid. Cause the liquids to rise in the glass tubes by sucking air out of the rubber tube. Pinch this tube, and measure the height of each liquid from the level of the liquid in its beaker. Determine the specific gravity as before.

EXERCISE 48. — *Specific Gravity Determined by Means of a Hydrometer.* Prepare a stick of pine 1 cm. square and 25 cm. long. Bore a hole in one end and force bullets into it until the rod will float in water in a vertical position and with a few cm. projecting above the surface. Coat with hot paraffine to make it waterproof. If, before coating the rod, the upper part of it is graduated in mm., it will facilitate the work with the instrument.

Float this hydrometer in water in a suitable vessel, note the depth to which it sinks. How much water does it displace? How much ought it to weigh? Now float in some other liquid, *e. g.*, salt water. Note the depth to which it sinks, and from your two measurements determine the specific gravity of the liquid tested.

PNEUMATICS.

EXERCISE 49. — Fit to a two-liter bottle a rubber stopper through which passes a short glass tube. The exact volume of the bottle should be determined once for all and marked on the bottle; this may be done by weighing it when empty and again when full of water.

Connect this bottle, by means of a thick-walled rubber tube, to an air-pump having a pressure gauge. Be careful to make all joints air-tight. Glycerine is useful for this purpose. Exhaust the air from the bottle. Close the rubber tube with a pinch-cock, and at the same time record the reading of the gauge. Now weigh bottle, stopper, tube, and pinch-cock; then open the pinch-cock, admitting the air, and again weigh. The difference is the weight of the air that was removed. Its volume will be found by comparing the reading of the pressure gauge with that of the barometer at the time of the experiment. From these data determine the specific gravity of air.

EXERCISE 50. — (a) Connect to one arm of a Y tube, by means of a perforated stopper, a glass tube 50 cm. long and about 15 mm. in diameter; to the other arm, a tube about 5 mm. in diameter and 50 cm. long. Over the stem of the Y slip a short rubber tube. Introduce both glass tubes into a vessel of water. Suck the air out of the rubber tube. What causes the water to rise in the glass tubes? How do the heights of the liquid in the two tubes compare?

(b) Measure the height of the mercury in a barometer tube. What supports the mercury? What change would it make in the height of the mercury column, if the cross-section of the column was 1 scm.? What would be the weight of such a column, if the specific gravity of mercury is 13.6?

What, then, by your calculation, is the pressure of air per scm.?

EXERCISE 51. — Work the lifting pump (or force pump), noting carefully the position of the valves. Make two diagrams of the pump, one showing the position of the valves during the *up* stroke of the piston, the other, during the *down* stroke of the piston.

EXERCISE 52. — Make a siphon of a rubber tube. Compare the velocity of water when the outer end is 2 cm. lower than the surface of water in the vessel with the velocity when the same end is 20 cm. lower.

EXERCISE 53. — Pour a little mercury into a Mariotte's tube, and adjust by shaking until the mercury stands the same height in both arms. It is evident that the pressure on the confined air is the same as external air pressure, or one atmosphere. Also, as the expansive force of an elastic body is equal to the pressure it supports, the expansive force of the confined air is one atmosphere. This may be expressed in $g.$, or, since air pressure is determined by the height of the barometer, we may represent the pressure and expansive force of the confined air by this height. Measure the length of the column of air in the short arm. As the tube should be of uniform diameter, this length may represent the volume of the confined air. Pour mercury into the long arm until the vertical distance between the mercury surfaces in the two arms is about 15 cm. Carefully measure this distance and the height (volume) of the air in the short arm. Now pour in mercury and measure as before until you have made five or six measurements, or until the tube is full. Record your results in tabulated form, and by comparing them, determine the law connecting the variation of pressure with variation of volume. (See Exercise 21. Note 1.)

HEAT.

In the work in heat it will save much of the pupils' time, if a vessel is provided for holding water that may be warmed to the temperature of the room, from which pupils may take water for experiments when water is needed at that temperature. Also provide a reservoir for hot water so time will not be taken by the pupils for heating it. If access can be had to a steam-pipe with a faucet, the exhaust steam will heat water very quickly.

EXERCISE 54. — Half fill a 4-ounce bottle with water or mercury. Carefully note its temperature, which should be the same as that of the room. Close the bottle with a stopper; wrap in several thicknesses of paper to protect from the heat of the hand, and shake vigorously. Note change of temperature. Shake again for a longer time, and again note the temperature. Evidently the change has been caused by the shaking. What transformation of energy is this? Give three other examples of such transformation.

EXERCISE 55. — Fit to the neck of a 4-ounce bottle a stopper having two holes. Half fill the bottle with water. Pass through one hole of the stopper the stem of an air thermometer bulb. Push it down until the end of the tube is well under water. Now clasp the bulb with both hands. Explain what occurs. Removing the hands, bring a flame near the bulb. Note the effect. Let the bulb cool; then apply cold water, — if possible, ice. What effect? Explain. Are two holes in the stopper necessary? Why?

EXERCISE 56. — Fill an 8-ounce flask (wide test tube) with cold water. Fit closely to its neck a stopper through which passes a tube 30 cm. long, having a bore of 1 or 2 mm. The water should stand in the tube 1 or 2 cm. above the stopper. Clasp the flask closely with both hands. Note effect on the height of the water in the tube. Next apply carefully the flame of a Bunsen burner. Note as before. If you watch closely, you may notice that *just* as the flame is applied, the water sinks. Can you account for this? What general effect of heat upon water does this experiment show?

EXERCISE 57. — *To Find the Coefficient of Linear Expansion of a Solid:* Fit up apparatus as indicated in diagram. This consists (1) of a tin tube 60 cm. long, 2.5 cm. in diameter, open at both ends. On opposite sides of this tube, near each end, is a short tube about .6 cm. in diameter. Near the middle is a tube about 1.5 cm. in diameter, into which a thermometer may be fitted. Perforated corks are fitted to each end of the tube. Through these passes a brass tube of about .6 cm. diameter, and a little longer than the large tube. . The whole is mounted so that one end of the brass tube presses against a screw at the right, the other against the short arm of a lever whose long arm moves before a ruler

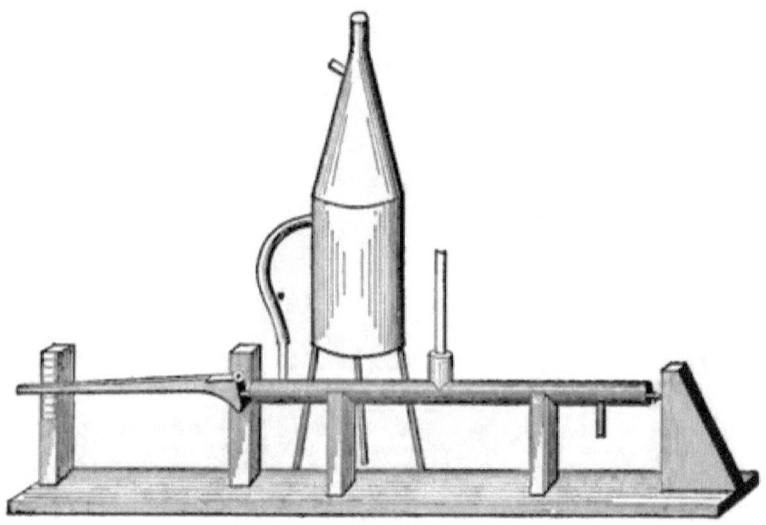

Fig. 11.

fastened vertically. (2) The other piece of apparatus is a boiler for generating steam. It is a sheet-copper cylindrical vessel 15 cm. tall and 10 cm. in diameter, mounted on three legs which raise it about 20 cm. The legs are kept from spreading by a flat ring of sheet metal connecting them at the bottom. Leading out from near the top is a tube 5 cm. long and 6 or 7 mm. in diameter, through which steam may be carried off in a slightly ascending direction when the top of the vessel is closed. A conical top 30 cm. tall fits the cylindrical vessel as a cover fits a pail. The joint must be as near steam-tight as possible so the top of the cylinder is not wired, but is left flexible. The open top of the conical

portion is about 2.5 cm. in diameter, and of such shape as to be easily closed with a stopper. A side tube, similar to the one on the cylindrical portion, leads out from near the top of the cone. These two admirable pieces, called the "Linear Expansion App." and "App. A," are described in Hall & Bergen, and are supplied by most dealers in laboratory materials.

Measure the brass rod when cold, and note reading of thermometer. Adjust the screw so that one edge only of the brass tube touches the lever. To do this easily it will be well to have the end of the tube beveled slightly. Now boil the water in App. A vigorously, and pass the steam through the tin tube. The two upper openings of App. A are of course closed. Note carefully the rise of the lever, reading to quarters, or less, of mm. When the lever ceases to rise, note temperature of steam, *i. e.*, of the brass tube. Measure carefully the lengths of the two arms of the lever. From this you can compute the amount of expansion of the tube. Now determine the ratio of expansion for one degree C. to the original length of the rod. This is the coefficient of expansion.

EXERCISE 58. — Fasten on a board, as indicated, wires 20 cm. long, of copper, brass, German silver, iron, etc. Heat at A with a Bunsen flame. Now beginning at the ends of the wire farthest from the flame, slide along each in turn the head of a match. Note the point where the match is ignited. Measure the distance from this point to the flame, and so make a list of the metals used, in the order of their abilities to conduct heat. Or: Move the tip of a finger carefully along each wire toward the flame until a point is reached which is uncomfortably warm. Measure the distance from this point to the flame, and thus make a list of the metals as above.

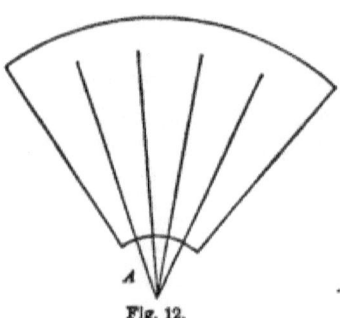

Fig. 12.

EXERCISE 59. — Fill a test tube nearly full of water. Incline it and apply the tip of a flame to it near the top. Take care that the flame does not touch any part of the glass not covered inside by the water. Hold in this position until water at the top of the tube boils. How is the temperature of the lower end of the tube

affected? What does this show as to the ability of water to conduct heat?

EXERCISE 60. — Fill a large beaker with water. Stir into it a little fine sawdust. Now gently heat the beaker at the bottom, close to one side. Watch for evidence of currents. Explain the cause of these currents from your results in Exercise 56.

Make a rectangle of glass tubing 10 cm. × 15 cm., joining the ends by using rubber tubes and a Y tube. Fill with water in which has been stirred a little fine sawdust. Cautiously heat one of the lower angles of the rectangle. Note and explain. Cool one of the upper angles by touching with a piece of ice. What occurs?

This illustrates how water circulates in buildings warmed by hot water.

Where would you apply heat to water to warm it most quickly?

In what way does the method of heating a metal differ from that of heating water? The first is called *Conduction;* the second, *Convection.*

EXERCISE 61. — "Touch paper" is made by dipping strips of unsized paper in a solution of saltpetre (potassium nitrate), the paper being dried before using. It burns with no flame and a great deal of smoke. Light a strip of this paper, and by means of the movements of the smoke, study currents of air. Recollect that smoke is made up of solid particles which, being visible and very light, serve to indicate the motions of the air. Why does the smoke rise? Hold it under the mouth of an inverted battery jar. Note, and explain the motion of the smoke. Light a piece of candle about 3 cm. long; set it on the table, and set over it a student-lamp chimney. Procure a strip of tin that will slip easily into the chimney. Near the end of this punch two holes, through which run a stout wire a little longer than the diameter of the chimney. Drop the strip down the chimney. This will divide the chimney into two passages. The candle should stand under one of the passages. Hold the burning paper close above the top of the other. Note and explain, as before, the movements of the smoke.

EXERCISE 62. — Fill a beaker, or other vessel, with cracked ice that has been washed clean. Insert in it a thermometer until the point marked 0° (the instrument is supposed to have the Centigrade scale) is just above the surface of the ice. When the mer-

cury column has come to rest, record the reading. In reading the thermometer in this and other experiments, be careful to stand in such a position that a line from the centre of the eye to the top of the mercury column will strike the stem of the thermometer at right angles. Read to quarters, and, if possible, to tenths of the smallest division on your instrument. On these two precautions will depend much of your success in the succeeding work. Now sprinkle on the ice a little salt. What effect? This shows the importance of having clean ice.

EXERCISE 63. — Fill App. A with water to the depth of 3 cm. Close the side tube of the cylinder. Fit a stopper to the top of the cone. Pass through it a thermometer, pushing it down until the point marked 100° is about 2 cm. above the stopper. The bulb, however, must not come within less than 2 or 3 cm. of the water in the vessel. Boil the water several minutes, read very carefully your thermometer, and record. In order to find what would be the boiling point of your thermometer under standard conditions, consult the barometer. The boiling point will be raised or lowered one degree for every excess or deficiency of 27 mm. from the standard 760 mm. What would be the boiling point of your thermometer under standard pressure? Do not boil the water violently lest the steam be compressed. Partly close the escape-pipe a moment with a cloth. Note the effect on the boiling point. Now lower the bulb into the water and note reading. Dissolve in the water some salt. What effect on the boiling point?

EXERCISE 64. — Fit to a wide test tube a stopper. Through it pass a short glass tube. Over the outer end of this slip a rubber tube about 10 cm. long. If it does not fit the tube tightly, wrap it with soft cord. Boil the water in the tube a few seconds; then, first removing the flame, pinch the rubber tube, invert the test tube and pour on it cold water. This condenses the water vapor in the tube and so lessens the pressure. In the preceding exercise you noticed how an increase of pressure affected the boiling point. What do you now find to be the effect of lowering the pressure?

EXERCISE 65. — *Boiling Point of Other Liquids.* Fit to a wide test tube a stopper having two holes. Through one pass a thermometer, and through the other a short glass tube the outer end of which is bent at right angles. Fill the test tube one-third full

of turpentine. Cork it and press the thermometer down into the liquid. Boil by means of a sand bath. This may be made by filling a tin pan 3 or 4 cm. deep by 10 or 15 cm. in diameter, with fine sand. Immerse the lower end of the test tube in the sand and apply heat from below. Boil and record temperature. Repeat with ether. But apply heat by introducing the test tube into a vessel of water heated to about 70° C. The flame must first be extinguished to avoid danger of explosion.

EXERCISE 66. — Fill a beaker with water. Record its temperature and that of the salt to be used. Stir into the water a handful of the salt, — better, powdered ammonium chloride. Again note temperature. What effect, if any, would the stirring have on the temperature? What, then, must have produced the change noted?

Mix snow and salt in portions of about 2 to 1 (the portions are not very important). Record the temperature. In the first part of this exercise the fall in temperature was due to the dissolving of the salt. Is this a similar phenomenon? You can decide this by noting which melts the faster, the mixed ice and salt, or the unused ice or snow.

Pour on the back of your hand a few drops of ether. Then, after this has evaporated, a few drops of water of about the same temperature as the ether. Which feels the colder? Which evaporates the faster? As both were at the same temperature, to what may the difference in sensations be due? Repeat by pouring a little of each liquid in turn on the air thermometer of Exercise 55.

State two other examples of cooling by evaporation.

EXERCISE 67. — *Apparatus:* A calorimeter; this must be a vessel of thin metal about 6 cm. in diameter and 12 cm. tall, brightly polished; a nickel-plated "lemonade shaker." Into this pour water to the depth of a few cm. Cool this gradually by adding to it ice or snow, — finally salt, if necessary. Stir the water continually with a thermometer that it may have a uniform temperature. Watch the outside of the bottom of the vessel for the appearance of mist. Note the highest temperature of the water (this is considered the temperature of the metal, and hence of the air in immediate contact with it) at which the mist appears. Then gradually warm the vessel. This may be done by pouring into it a few drops of warm water.

Note the temperature at which the dew unmistakably begins to disappear. The average of these two readings gives the dew point of the atmosphere. Record the temperature of the room, the temperature out of doors, and the kind of weather. How much must the temperature fall to-day in order that it may rain?

EXERCISE 68. — Determine how the rate at which a body cools is affected by the temperature of the surrounding medium, *e. g.*, the air. Select for the body whose cooling is to be watched a thermometer bulb. One with a large bulb is preferable.

Remove the top from a tin can of about one liter capacity. Solder to this can three strips of tin by which it may be tacked to a board covering the top. Through this cover bore a hole,— about ⅜ in. in diameter. Blacken the inside of the can and the under side of the cover. This may be done by painting with thin shellac varnish mixed with dry lamp black. Immerse the can in a large vessel of water. Note temperature of the water, *which should be the same as the temperature of the room.* Slide the stem of the thermometer through a stopper that fits the hole in the board. Warm the thermometer to about 80° C. by holding it above (not in) a flame. Dry the bulb, if it has become moist. Now introduce the thermometer into the hole in the board and adjust so the bulb shall be near the centre of the can.

The temperature of the air now around the bulb is assumed to be that of the water, which you have noted. This will remain practically constant, since the blackened can will transmit rapidly to the water any heat received by the air from the thermometer, and the large body of water will not have its temperature perceptibly raised. Note temperature of thermometer when *finally* adjusted, and at the end of each succeeding half-minute. Record results as below:

(1) TEMPERATURE OF AIR IN CAN.	(2) TEMPERATURE OF THERMOMETER.	(3) FALL OF TEMP. OF THERMOMETER.	DIFFERENCE BETWEEN 1 & 2.
....	....	....	....

State, by studying last two columns, a rule governing the fall in temperature of a body. (Page 20, Note 1.)

EXERCISE 69. — Fill a beaker about one-third full of water, and add about an equal amount of pounded ice or snow. Apply the heat of a Bunsen flame. Stirring rapidly with a thermometer it will be noticed that the ice melts, but that the temperature remains practically the same as long as there is any considerable portion of ice. Since the vessel is receiving heat from the flame and the surrounding air, the heat so received is said to become latent. That is, it is used in so altering the cohesion of ice that it becomes liquid. Determine the amount of heat that becomes latent in turning 1 g. of ice at 0° to water at 0°. For our unit of heat we will take the amount of heat needed to warm 1 g. of water 1° C. A heat unit will, of course, be given up by 1 g. of water in cooling 1° C.

Weigh a dry beaker of, say, 300 ccm. capacity. Pour into it about 200 g. of water. Warm the water to about 50° and accurately take the temperature. Drop into the water small pieces of clear dry ice until about 100 g. have been added, i. e., until the level of the water is raised about one-half. Stir constantly, but not violently. Note the temperature when the last piece of ice is melted. If the ice has not all melted by the time the water has been cooled to 5°, dip out the remaining ice, taking as little water as possible. It is also important that the final temperature be about as much below the temperature of the room as the temperature at the first was above it, that the influence of the surrounding air may be disregarded. If the ice put in is not enough to do this, more may be added. Re-weigh to determine the exact amount of ice melted. From your results calculate the number of heat units necessary to melt 1 g. of ice. In doing this, *take notice* that the ice is not only melted, but also warmed to the final temperature, and that the amount of heat used in both these operations is equal to that given up by the beaker and the water ; — the amount of heat given up by a g. of glass in cooling one degree may be taken as .2 of a heat unit. If a metallic vessel is used, of course *its* specific heat will be taken.

Fig. 13.

EXERCISE 70. — *Determine the Latent Heat of Water Vapor.* Fit to the lower delivery tube of App. A (all the other openings being closed) an arrangement indicated in the diagram. The delivery tube should reach nearly to the bottom of a beaker

containing, say, 250 g. of water at about 15° below the temperature of the room. Accurately take its temperature. Boil water in App. A. Wait until all the delivery tube is hot and a strong jet of steam is issuing from it; then introduce into the water of the beaker, bringing the end near to the bottom. Condense enough steam to raise the water to about the same number of degrees above the temperature of the room as it was at first below it. Why? Then remove the delivery tube. Stir the water thoroughly and note its temperature very carefully. Again weigh and determine the weight of steam condensed. Now calculate the heat given up by a g. of steam in condensing to water at 100°. In working this problem notice that the heat which warms the water and the beaker comes from two sources, — the condensation of the steam to water at 100°, and the cooling of the water thus formed from 100° to the final temperature. Allow for the heating of the beaker as in the previous exercise.

EXERCISE 71. — In our last two exercises we have made use of the fact that glass in cooling furnishes only .2 as much heat as the same weight of water; or, the specific heat of glass is .2.

Determine the specific heat of shot. Place in the dipper of App. A 500 g. of fine shot. Cover the dipper with a piece of cardboard in which is a small hole through which a thermometer may be passed. Pour into a calorimeter 100 g. of water cooled to 6° or 8° below the temperature of the room. Place the dipper into App. A, which should be filled with water nearly to the level of the bottom of the dipper. Close the side tube of the App. Boil the water, stirring the shot continually. When the shot has reached the temperature of 100°, pour it quickly into the calorimeter. Stir thoroughly and when sure the temperature no longer rises, record reading of thermometer.

By a similar method to the one used in the two previous experiments, calculate the heat yielded by 1 g. of shot cooling 1°, noticing that in this case the vessel containing the water is of metal and its specific heat, if brass, is about .09. If the vessel is only partly filled, consider that only so much as is below the water line is warmed.

ELECTRICAL ENERGY FROM FRICTION.

EXERCISE 72. — Rub one end of a warm glass tube with a warm piece of silk, and hold the rubbed end near some bits of tissue paper and some small pieces of pith. Note carefully what follows. Repeat several times to be sure of results. Now rub a piece of sealing-wax with dry, warm flannel and do as you did with the glass rod. Results? The peculiar action manifested by these rubbed bodies is due to electrification produced by the rubbing. Did you do work in producing electrification? The kind developed on the glass rod is called positive electrification, and that on the sealing-wax, negative electrification. Bodies exhibiting these phenomena are said to have charges of electrification upon them.

Make a paper stirrup large enough to allow the rod or the wax to pass into it easily, and suspend it by a silk thread. Rub the rod, and, without touching the rubbed end with the hand, place it in the stirrup. Now rub another glass rod and hold it near the rubbed end of the suspended rod. What result do you get? Try it several times. Do the same with two sticks of sealing-wax, and record results.

Now hold a rubbed glass rod near to a suspended stick of wax on which there is a charge of negative electrification, and record results. Hold a rubbed stick of wax near a suspended glass rod on which there is a charge of positive electrification, and record results. State the law of action between charges of electrification covering the phases shown by the above.

EXERCISE 73. — Suspend a pith-ball by a silk thread. Call this an electroscope. Develop positive electrification and hold the rubbed body near the pith-ball. Note fully the action of the ball. If the ball is attracted to the rod and then flies away from it, try to touch the ball with the rod, and account for the result. Now rub the stick of wax, and hold it near the ball. Are the results the same? Why? Rub the wax, hold it near the ball, and after the ball has touched it and swung away, bring a rubbed glass rod near the ball, and account for results. Hold an unelectrified, or neutral, body near a positively charged electroscope, and then near a negatively charged electroscope. What is the action in each case? If a body has electrification on it, how would you proceed to find its kind? Can electrification do work?

EXERCISE 74. — We have seen that certain bodies, upon coming in contact with an electrified body, become electrified. Mount two metallic shells or balls, such as are used on the ends of curtain rods (or two potatoes may be used) on some short sticks of sealing-wax so that they will be on about the same level. Some pith-ball electroscopes can be charged, — one with negative, and the other with positive electrification, to be ready for any tests you wish to make, remembering that the repellent action is the true test for the kind of charge on a body. Hold a rubbed glass rod near one of the shells, having placed the two shells in contact, and approach the shells from the opposite direction with the positively charged electroscope. Result? Repeat. Result? Slowly remove the glass rod and note the action of the electroscope. Test the two shells for electrification. Result? Touch the shells and the pith-ball with your hand. Now re-charge the pith-ball electroscope positively. Hold the electrified rod as before, being careful that no sparks pass to the shells, and be sure the shells touch, and again bring the pith-ball up as before. While the rod is held near the first shell, carefully remove the shell and the rod. Test the two shells for the kinds of electrification. Results? This process is called Charging by Induction.

EXERCISE 75. — Place the two insulated shells, having pith-ball electroscopes, about a foot apart, and connect them by a stick of wax, as shown in the sketch. Hold an electrified rod near or against one shell and note whether the wax conveys the electrification to the other shell. Try, as connectors, a glass rod, a ruler, a silk thread, a piece of rubber tubing, and a piece of metal, and then try a wet silk thread. Make a list of those substances which allow electrification to pass to the second shell. These are called *conductors;* those which do not allow the electrification to pass are called *non-conductors.* Name them. Why do the shells rest on wax?

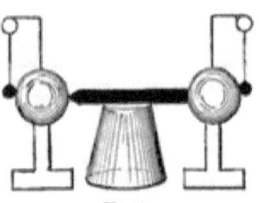

Fig 14.

EXERCISE 76. — Work an electric machine, and slowly separate the knobs in front of the wheels. Do you notice any difference in the frequency of the sparks when the knobs are near or far apart? If you get a large spark, it indicates a large charge of electrifica-

tion. Should you desire to increase the electrification on the machine, how would you have to affect the air space through which the spark passes from one knob to the other? In some respects this action of electrification very much resembles liquid pressure. If it is desired to have great water pressure throughout the city, the pipes through which the water passes must be strong; and if we desire to have a great charge of electrification on a body, we must increase the air space between that body and any other body to which the electrification might pass. The term used in regard to electrification, as pressure is in liquids, is *potential*. A body is said to be at high potential when it must be carefully insulated to keep its electrification from passing to the earth. Bring one knob of the electrical machine near an insulated shell provided with a pith-ball electroscope, and pass some tiny sparks onto it. Does the number of sparks increase the potential of the shell? The quantity of electrification on an insulated conductor is called its charge; what effect does increasing the charge of a given conductor have on its potential? Charge two insulated shells provided with pith-ball electroscopes with positive electrification so that they will have different potentials. Now bring them together and note the effect on the electroscopes. Repeat several times. This effect will always be seen when two bodies, at different potentials, are connected by a conductor. The transfer of electrification by means of a conductor is called a *current*. The earth is said to be zero potential. Positively charged bodies would send a current to the earth, and with negatively charged bodies, the current is said to be in the opposite direction.

EXERCISE 77. — One end of a narrow strip of board should rest on an insulated support, and the other on any good conductor, such as the steam-pipe. Some small screw-eyes can be screwed into the under side of the strip at equal distances, say 60 cm.; and from the screw-eyes suspend double electroscopes by means of No. 36 bare copper wire. The balls should all be of one size; the wires by which they are suspended should be the same length, and the ends of the wires should be tipped with wax. Fasten the wires with a loop at the upper end so that the balls will be free to swing sideways. Now pass sparks from an electrical machine onto the strip at the insulated end, and watch the balls. Does the electrification pass the whole length of the strip? Is it equally distributed

throughout the whole length? Does all of the electrification which starts from the insulated end reach the other end? Whatever tends to prevent the transference of electrification from one point to another, is called resistance. Does the strip offer resistance to the passage of electrification? Was electrification used up in overcoming this resistance? What is that which can overcome resistance and is used up in so doing? Does electrification seem to be a form of energy? What kind of energy is electrification when it is stationary? When it moves along a conductor? Are heat and light ever produced by the passage of electrical charges? What transformation of energy has taken place when a spark is seen?

EXERCISE 78. — Whatever tends to cause a transference of electrification from one place to another is called an Electro-Motive Force, — written E.M.F. A difference of potential between two points is an E.M.F. Can you think of any way of producing an E.M.F. without an expenditure of energy? If a difference of potential causes an E.M.F., does the E.M.F. probably increase with an increase of difference of potential? When you work the electrical machine, one discharging knob becomes positively, and the other negatively, charged. Is the E.M.F. greater or less between the two knobs than between either of them and an unelectrified body? Try it by holding an unelectrified body as far from one of the knobs as the knobs are apart, and note the action of the spark. Results?

EXERCISE 79. — In the preceding exercises you have found it necessary to insulate a body in order to electrify it. Now try to electrify an uninsulated body. A piece of paper, say 5 × 8 in., is fastened to one side of a pane of glass by mucilage on the corners of the paper. Warm the glass slightly and lay it, paper side down, on the table, and on the glass, just over the paper before used, place another similar paper. The two are now separated by the glass. Rub the upper paper with a cat-skin or piece of sheet rubber. Handle the glass carefully by the corners and bring the paper surfaces successively near a suspended pith-ball. Does either paper seem to be electrified? Hold the glass by one corner, and pull off the loose paper by the corner and test it to see whether it is more highly electrified now than it was when on the glass. Test the other paper now. Results? Repeat and test each paper for the kind of electrification on it, and note whether or not the

two have about equal amounts on them. Results? How did the lower paper become electrified? An instrument like this which enables us to accumulate quantities of electrification is called a condenser. The most common form consists of a glass jar instead of a glass plate, and instead of paper coats, it has tinfoil. The jars so made are called Leyden Jars, and the process of charging them with electrification is just the same, only, instead of rubbing either coat, the knob in the cork of the jar is connected to the inner coat of tinfoil by a little chain, and sparks from the prime conductor of an electrical machine are passed on to the knob, while the outer coat is connected with the ground, either by standing the jar on the table or by holding it in your hand. Do not try to get a very strong charge, and handle the jar very carefully. (You would better get the assistance of your instructor). Charge the jar thus, and having placed it on a glass plate, or cake of paraffine, connect the outer coating to the knob by means of a metallic connector having an insulator handle. What follows? Why? In a few seconds connect the coats again. Results?

If there was more of one kind of electrification than of the other, what will be the condition of the jar after the sparks cease? Test the jar by approaching it with an electroscope. Results?

CURRENT ELECTRICAL ENERGY.

EXERCISE 80. — In your previous work you learned that if two bodies were at different potentials and were connected by a conductor, there was a transference of electrification from one to the other. If the bodies could be kept at different potentials, the transference of electrification, or current, as it is called, would be continuous. Such devices have been found, and one of them is now to be studied.

Make a weak solution of sulphuric acid by pouring about 2 ccm. of the acid into a tumbler two-thirds full of water. Stand in this solution a brightly sand-papered strip of zinc 10 cm. by 4 cm. Note carefully for a few minutes any changes on the surface of the metal. Also stand a similar strip of bright sheet-copper in the solution, but not in contact with the zinc. Is there any marked change on its surface? Remove the zinc, and while the surface is still wet with the acid solution, spread a drop of mercury on the zinc. (Do not let your rings come in contact with the mercury.) A little cloth may be used to spread the mercury so that the entire surface of the zinc becomes bright like a mirror. Coating the zinc thus is called amalgamating it. Replace the zinc in the solution. Does the same action appear on its surface? Keep the bottoms of the metals apart, but lean the tops together, and watch the surfaces for the evolution of bubbles. Results?

EXERCISE 81. — To make it more convenient, you can use a strip of pine about 2 cm. square, and long enough to rest across the tumbler, and fasten the metals on opposite sides of the pine by driving through each strip a tack, around whose head is wound the bare end of a piece of No. 24 insulated copper wire. Having thus arranged the metals, stand them in the acid solution and press together the two free ends of the wire, being careful to have them scraped bright with a knife or sand-paper. Is any action manifested on either of the metal plates? Separate the wires Result? Repeat several times, and record results. Connecting the wires thus is called closing the circuit, and separating them is called breaking the circuit.

Examine two of the strips which have been used for some time, and tell which one of them the acid solution has acted on the more

vigorously. This "plate" is the one where the electrical energy is generated, and is called the *positive plate*. It is at a higher potential than the other, and the E.M.F. causes the current to pass from it through the liquid to the other, or negative plate, whence the current passes through the wire (external path) back to the first plate. Name in order the substances through which the current passes in making a complete circuit.

The free end of the wire connected to the *negative* plate is called the *positive electrode;* while the free end of the wire connected to the *positive* plate is called the *negative electrode*. A vessel containing a pair of metals arranged similarly to the one you have used, and having in it a solution which will act chemically on one of the metals, is called a *Galvanic* or *Voltaic* cell or element. When several cells are used and are connected with one another, they constitute a Galvanic battery. (A single cell is sometimes called a battery.)

Caution: Never twist the ends of wires when you make connections; use double wire couplers. When you are through with the pieces of wire, wind them carefully onto a spool so as to keep them from getting twisted or tangled.

EXERCISE 82. — There are many kinds of Galvanic cells, and among them are some two-fluid cells, one of which is made as follows : use for materials a strong glass tumbler, about 10 cm. tall, and 7 or 8 cm. wide ;. a cup of unglazed porcelain, about 10 cm. tall, and 4 or 5 cm. wide ; about half a liter of dilute sulphuric acid (20 parts in volume of water to 1 part in volume of concentrated acid); about half a liter of a saturated solution of copper sulphate; a piece of zinc 10 cm. long, 2.5 cm. wide, and .5 cm. thick, having about 50 cm. of No. 20 insulated copper wire fastened firmly to it; a piece of sheet-copper 10 cm. square, and having a similar wire fastened to it.

Put the zinc into the porous cup, and then pour in dilute acid until the cup is full to a level, 2 cm. below the top. Now put the cup into the tumbler and pour into the tumbler the sulphate of copper solution till the tumbler is full to a level with the acid in the porous cup. Remove the zinc and amalgamate it with mercury and replace it in the porous cup. Put the sheet-copper, bent into a circular form, into the sulphate of copper solution. Put the cell in circuit with a galvanometer, and note the deflection when the needle comes to rest, and then while the current continues to pass

through the galvanometer, note the deflections for periods of 5 min. each for 30 min., if time permits. Results? Compare with results obtained when a single fluid cell is so used. (You may need the aid of your instructor in using the galvanometer.)

EXERCISE 83. — Among the two-fluid cells the most common is the so-called Gravity cell. It consists of a zinc plate, a copper plate, a solution of zinc sulphate, and a solution of copper sulphate, a jar to hold the solutions and plates. The copper plate is placed in the bottom of the jar, and is covered with a saturated solution of copper sulphate; and from this plate is extended up out of the jar a copper wire covered with hard rubber. Then a solution of zinc sulphate is carefully poured in, and in a few minutes is seen to be resting on the copper sulphate solution. The zinc plate is now suspended from the top of the jar so as to be immersed in the light colored solution at the top of the jar, care being taken not to let it down into the blue solution; the zinc should be covered with liquid, — water may be added to get this result.

When the plates are connected by wires, some chemical changes take place in the molecules of the solutions. The copper sulphate solution deposits copper on the copper plate, and zinc from the zinc sulphate takes its place; and the zinc plate then gives off more zinc to produce new molecules of zinc sulphate. In time the copper sulphate will be all used up, and there will be too much zinc sulphate solution. This is shown by the liquid all becoming clear. Some of the clear solution must then be poured off, and a handful of crystals of copper sulphate be poured into the cell. Compare the strength of the current from this cell with that of the other two-fluid cell. Results? Why called a Gravity cell?

EXERCISE 84. — We cannot see the current when a battery is in action, but we can see some of its effects, and you are now to study some of them. Use a battery of two or three large cells, having them connected in series, *i. e.*, the positive plate of one cell connected to the negative plate of the next, and so on, and then between the copper wires which connect the last two plates place a short piece of very fine iron or platinum wire. Note carefully any change in color of this fine wire, or, if its color does not change, carefully touch it with the finger. Results?

Remove the fine wire and connect the two copper wires by a coupler; scrape the insulation from a short section of the wire through which the current is passing, and dip this bare section

into some iron filings. Result? Break the circuit, and note the effect on the filings. Repeat several times to be sure of what happens, and record results. What is here shown to be the condition of a wire carrying a current of electricity?

Now dip the free ends of the wires coming from the battery into a solution of potassic iodide (made by dissolving a few crystals of potassic iodide in water, and then stirring it into a thin starch paste), and note the effect at the places where the wires touch the solution. A dark or purple color indicates a chemical change by which iodine is set free. Do you get this result? Name the three effects of current electricity you have seen in this exercise.

EXERCISE 85. — Wind some insulated copper wire No. 24 around a piece of soft iron, such as a wrought iron nail. The wire should be about 1 m. long, and should have about 30 cm. free at each end after winding onto the iron so as to give freedom of movement after connecting the free ends of the wires to a galvanic cell. (The tumbler cell will do.)

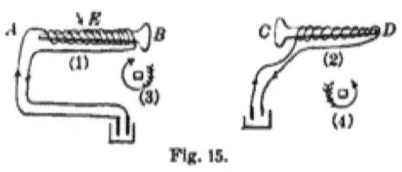

Fig. 15.

Wind two as shown in Fig. 15. The figures show the wire on the side next to you as you wind them. Having connected the ends of (1) as shown above, dip A and B successively into a dish of iron filings. Results? Approach the north-seeking pole of a magnetoscope with A and then with B. Results? What is the nail now, and what can you say about end A and end B? Now disconnect from the cell, and again dip the ends of the nail into filings. Results? Under what conditions is a piece of soft iron a magnet? If you imagine yourself swimming in the "current of electricity" in any section of the wire, as at E, and facing the soft iron, on which side of you is the north-seeking pole of the magnet which the soft iron becomes? Now make the same trials, and answer the questions, using 2, and record your results.

These devices are called electro-magnets while the current passes through the insulated wires. The coil is called a helix, and the soft iron, its core. If you look at the end of 1, as at A, where the current enters the helix, the current will pass around the core, as is shown in 3. Compare this direction with the direction of motion of the hands of a clock. This is a right-hand helix.

Looking at 2 where the current enters at *C*, the current will travel around the core, as is shown in 4. Compare this direction with the direction of the motion of the hands of a clock. Frame a rule for telling the poles of an electro-magnet when you know the direction of the current around it.

If you were given an electro-magnet and could not see the battery to which it is connected, how could you tell the direction of the current in the wire?

EXERCISE 86. — You will recall the effect on a freely suspended magnet when another magnet was brought near it. From what you saw, when a wire carrying a current was dipped into filings, what do you think would be the effect on a freely poised magnet, should a wire carrying a current be held near it? Use an insulated copper wire, No. 24, about 1 m. long. Connect the bare ends of the wire to the poles of a cell and, remembering that the current is from the negative plate through the wire to the positive plate (from copper to zinc), hold the wire so that a section of it will carry the current from the north to the south near a delicately poised magnet, and get results from each of the following conditions: (Do not turn the wire, even though the needle should turn when the wire approaches it.)

Current going from	{	1. North to south above magnet.	}	North-seeking pole deflected towards east or west?	{	1. ....
		2. South to north below "				2. ....
		3. South to north above "				3. ....
		4. North to south below "				4. ....

Examine carefully the above results, and, thinking of yourself as swimming in the "current of electricity" and always facing the magnet, on which side of you will you find the north-seeking pole of the magnet deflected? State the rule clearly and briefly.

EXERCISE 87. — If the wire should be held so that the current would pass from north to south above a freely-poised magnet, the

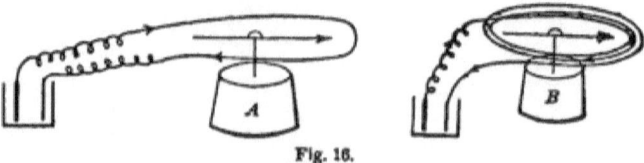

Fig. 16.

north-seeking pole would be deflected in what direction? If from

south to north below, in what direction? If now you were to hold the wire so that one section of it would carry the current from north to south above, and another section, from south to north below, how would the amount of deflection of the north-seeking pole compare with either of the first-mentioned cases above? Try it (*A*). Results? Try by having several coils, as shown in *B*, and compare results with the trial in *A*, and state any inferences.

Use a larger battery cell, and compare results. Can you think of any use to which you could put an instrument of this kind?

Instruments having several coils of wire around a delicately-poised magnet, under which is a scale graduated in degrees, is called a galvanometer. How would increasing the number of coils (other things being the same) affect the deflection? If the number of coils remained constant, how would increasing the current affect the deflection?

EXERCISE 88. — A tangent galvanometer is now to be used. It should stand on the table so that the coils and the magnet are parallel to each other; this will be true when the pointers are on the zero marks. The instrument should be level so that the needle is in the center of the coils, and free to move. There should be no iron or magnets near, and if the needle tends to stick, tap the base gently before taking a reading. Be careful not to change the coils from their north and south position while making tests. The earth tends to keep the needle in a north and south position, while a current passing through the coils tends to turn the needle east and west, so that when the galvanometer is in use, the needle assumes a position which is the resultant effect of the two forces acting upon it. (Some galvanometers have what is called an *astatic needle*, *i. e.*, a combination of two magnets, one being placed above the other, with opposite poles near each other, so that the earth's effect on one magnet neutralizes the effect of the other, and the influence of the earth is thus eliminated. Such galvanometers are much more sensitive than the tangent galvanometers.)

If the number of coils used in the galvanometer be constant, a greater deflection of the needle indicates what? Now find the deflections produced by two or three different cells, and record the results. It is better to read by reversals, *i. e.*, cause the current to pass through the coils in one direction, and then, after noting the

deflection, cause the current to pass in the opposite direction, and note the deflection. The average of the two is the true reading. (To reverse the current in the galvanometer, a commutator may be used, or, have two wires, about 30 cm. long, coming from the galvanometer, and on the free ends have wire couplers, so that connections may be made without disturbing the instrument.) Now find the deflections produced by two or three different cells, and record results; always get readings accurately to one-fourth of a degree. The strengths of currents are not proportional to the degrees of deflection, but are proportional to the tangents of the degrees, *i. e.*, if one cell gives a deflection of 55°, and another cell gives a deflection of 35°, the strengths of the currents are not as 55 : 35, but as tangent 55 to tangent 35. The table of natural tangents is found on p. 150, and by consulting it you will find tangent 55 to be 1.428, and tangent 35 to be .700, so that the strengths of the currents are to each other as 1.428 : .700, or about as 2 : 1.

EXERCISE 89. — Connect one plate of a cell with one post of a galvanometer, and from the other plate of the cell run a wire about 30 cm. long, having its free end scraped bright. Connect a similar wire to the other post of the galvanometer. Now press these two free bright ends of the wire against or into each of the following substances, and note the amount of deflection made in each case, so as to determine which substances allow the current to pass through them easily : — iron, brass, copper, wood, glass, carbon, silver, paper, pure water, dilute sulphuric acid and zinc. Tabulate as follows : —

SUBSTANCE.	GALVANOMETER DEFLECTION.

By examining the list and the deflections, number the substances in the order of the *power to resist* a current passing through them.

Why is copper wire generally used? Why is it covered with cotton or silk? Why is it necessary to remove the insulation in making connections?

EXERCISE 90. — You have seen that substances differ in their ability to resist the passage of an electrical current through them. To find just how much resistance a certain object offers makes it necessary to have some standard of measurement. The unit accepted is called the "ohm," and is the resistance of a column of mercury 106.3 cm. in height, and of uniform cross-section, and having a weight of 14.4521 g. at a temperature of 0° C. Instruments consisting of several coils of wires of different known resistances varying from .1 to 500, or more, ohms are now prepared, and by using such an instrument together with a galvanometer, you can determine the exact amount of electrical resistance of any conductor. The resistance box and galvanometer are connected to a battery or cell so that the current passes through all of them, thus: Fig. 17. The keys of the resistance box are all placed on the buttons marked zero, and the deflection of the needle noted. Now introduce more and more of the coils into the circuit by moving the keys around on successive buttons, thus causing the current to meet with greater and greater resistance to its passage. How does this affect the current, and what effect does it have on the needle? Now, with any resistance, say 5 ohms, read the deflection. Break the circuit and then place a bar magnet on the table about 20 or 30 cm. from the coil, and have it exactly in line with the needle when its pointer is on the zero mark, having the north-seeking pole of the magnet towards the south-seeking pole of the needle. Close the circuit and again note the reading. Does the presence of the magnet increase or decrease the sensitiveness of the galvanometer? Reverse the ends of the bar magnet, and find the effect upon the readings. Vary the distance of the bar magnet, and record results.

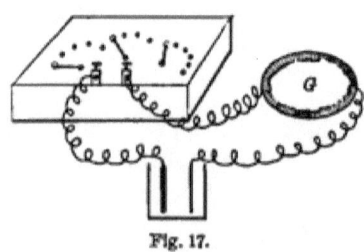

Fig. 17.

EXERCISE 91. — One method of measuring resistances is as follows: The object whose resistance is sought is placed in circuit with a battery and a galvanometer, and the deflection noted. (Use a magnet and reduce the reading to about 45°.) It is better to read by reversals, and take the average reading of several trials. The object is then removed, and a resistance box is substituted in

its place and resistances are introduced in the box till the galvanometer gives the same deflection as was obtained when the former reading was taken (read to quarters of degrees), when the total resistance introduced by the box will be the resistance of the object tested. Record your results. This is called determining resistance by substitution. By the process just described, determine the resistance of several coils of wire and record results as follows:

	No. of Wire, B. & S.	Diameter.	Length.	Kind of Ware.	Resistance.
1	22	.644 mm.	10 m.	Copper.	
2	28	.321 mm.	10 m.	"	
3	22		20 m.	"	
4	28		20 m.	"	
5	28		10 m.	German silver.	

From the above table and the results you obtained, can you see a definite relation existing between the lengths of conductors of the same diameters and their resistances? State it. How do resistances vary in regard to diameters or sectional areas of wires of the same length?

The results from (2) and (5) indicate what in regard to material?

EXERCISE 92. — To measure the resistance of a battery by means of a tangent galvanometer.

The battery is connected to the galvanometer so that the deflection is about 45°, and the exact deflection is noted. Now by means of a resistance box introduce resistance till an angle is obtained whose tangent is just half of the tangent of the angle of the former deflection. (Consult the table of angles and their tangents on p. 151.) The resistances offered by two bodies to a current are to each other inversely as the tangents of the angles of deflection on the tangent galvanometer, hence the resistance introduced by the box above is equal to the resistance of the *coils* used in the galvanometer and the resistance of the battery.

If now you deduct the resistance of the coils used in the galvanometer from the resistance introduced in the box, you will have the resistance of the battery. For example, if the battery and galvanometer in circuit give a deflection of 45° the tangent of

45° is 1.000. Half that tangent is .500. Tangent .500 is that of angle 27°. The resistance introduced to bring the deflection to 27° is 8.2 ohms. The resistance of the galvanometer coils used is, say, 1.5 ohms. Then $8.2 - 1.5 = 6.7$; therefore, the resistance of the battery is 6.7 ohms. Measure the resistance of two different cells thus and record your results.

EXERCISE 93. — Use an *astatic galvanometer* and a Wheatstone Bridge. The Wheatstone Bridge presents one of the best known

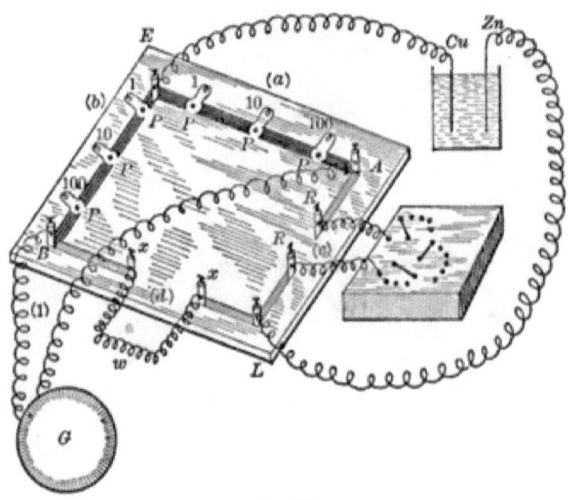

Fig. 18.

methods of quickly and accurately measuring resistances. It works on the following principle: The current starting from the battery reaches the point marked E on a thick square of wire or a strip of metal; here it divides, and if R and R are connected by a thick copper wire, and X and X are similarly connected, then equal amounts of the current pass over E, A, L and E, B, L, the two amounts uniting at L and passing on to the battery. For every point in E, B, L there is a point in E, A, L having the same potential, hence should any two such points be connected, there would be no current through the connecting medium. A and B are two such points, being equally distant from E, and should you insert a galvanometer between them, there would be no deflection of its needle. If, now, you should introduce a coil of wire between X and X, one division of the current would have to pass through

the path E, B, w, L, and would, therefore, meet more resistance than it would by passing over the path E, B, G, A, L, so that part of the current would pass through the galvanometer, as shown at 1. If you should remove the heavy wire between R and R, and put it between X and X, and should put the coil w between R and R, the process would be reversed, and the division of the current going over the path E, A, L would subdivide at A, part going over E, A, G, B, L, thus giving a reverse deflection in the galvanometer. Use No. 16 or 18 wires of about equal lengths to make connections, and set up the apparatus as shown in the diagram. Have two keys for closing and breaking the circuits, one between the point L and the battery, and the other between A and the galvanometer. (Always close the battery circuit first.) Close the circuits and adjust the switches in the resistance box till there is no deflection of the galvanometer needle. The resistances of the wire coil and that introduced in the box are now equal. What is the resistance of the wire tested? Test several different coils, and record results.

It is better to unplug unequal resistances in the arms of the bridge marked (a) and (b); for example 10 ohms in (a) and 1 ohm in (b) if the object to be tested is one of *small* resistance, and if it is one of great resistance unplug 10 ohms in one arm and 100 ohms in the other. Be sure that the plugs marked (P) are in firm contact with the metal strip unless the resistances marked on their spools are to be thrown into the circuit.

Close the circuits and adjust the switches in the resistance box till no deflection of the galvanometer needle can be noticed. Then since it is the principle of the Wheatstone Bridge that when the above condition is obtained the products of the resistances of the opposite arms are equal, *e.g.*, $a:b::c:d$ or $d = \frac{bc}{a}$, you can find the numerical value of (d) which will be the resistance sought.

Find the resistances of several coils and record results.

EXERCISE 94. Wind several m. of insulated copper wire, No. 18 or 20, around a pasteboard cylinder about 4 cm. in diameter, and leave about 2 m. of each end of the wire unwound so as to make connections with an astatic galvanometer. Place some books on the wires near the galvanometer, so that in moving the coil you will not move or disturb the galvanometer, and in your manipulation keep the coil as far from the galvanometer as possible. Slide

the coil suddenly over one end of a bar magnet, as you would a ring on a finger, and note the effect on the needles of the galvanometer. Remove the coil suddenly, and note the effect. Repeat several times, to be sure of results. Now reverse the ends of the magnet, and do as before, and note how the deflections compare with the former ones. How does the direction of the deflection produced by sliding the coil *onto* one end of the magnet, compare with that of the deflection made by removing it from the other end of the magnet? Make several trials. Result?

Hold the coil *still* while it surrounds the magnet, and note whether a current is manifested in the galvanometer. When the coil is moving on or off the magnet, the number of lines of magnetic force passing through the closed circuit of the wire is varied, *i. e.*, when the coil passes onto the magnet, the number of lines of magnetic force passing through the closed circuit of wire is being increased, and while the coil is being withdrawn, the number of lines of force is being diminished. Will either process cause a current? When the number is not being varied, is there a current? How does the deflection made by increasing the number of lines at one pole of the magnet, compare with decreasing the number of lines at the other pole?

How could you arrange a combination of this kind to make a continuous current in one direction?

This is the plan of a dynamo. If one is at hand, examine with the aid of your instructor.

EXERCISE 95. — Wind about half a meter of No. 30 insulated copper wire around a wrought iron nail, and leave about 1 m. of each end of the wire free. Suspend the nail by its "point," as shown in Fig. 19, connecting the wires to a battery. Make a similar electromagnet which you can use without suspending it. It can, however, rest on a cylinder so as to be near *B*. Cause the current to pass through *A* successively in opposite directions by changing the connections at *P*

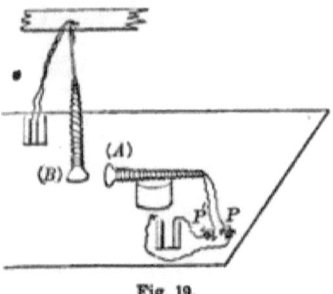

Fig. 19.

and *P'*. (Simply press the bare ends of the wires from *A* onto the posts *P* and *P'* and time the connections so as to be in unison,

if possible, with the motion of *B*, if *B* moves.) Do you get motion without touching *B* or breathing against it? Why?

If you could have some electro-magnets arranged on an axle so that their ends would project like the spokes of a wheel from the

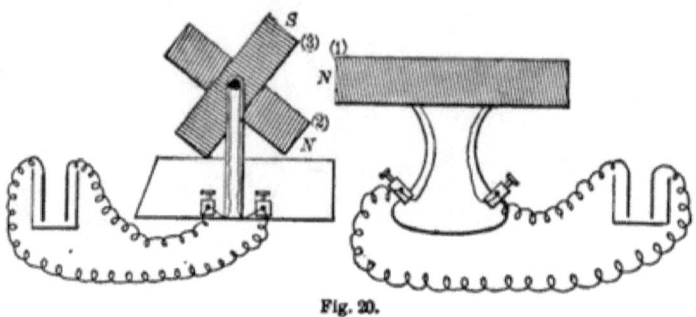

Fig. 20.

hub, as shown in Fig. 20, and have the poles as indicated, what would the joint action of (1) on (2) and (1) on (3) cause the wheel to do? If you could continually reverse the currents in the "wheel" electro-magnets so as to keep the poles in the vicinity of 1 of the polarity indicated, what would happen to the wheel? This is the plan of an electro-motor, and the connections with the wheel coils are made through the axle.

SOUND.

EXERCISE 96. — Suspend a solid rubber cord 4 or 5 mm. in diameter, and 4 or 5 m. long, from a hook in the ceiling, or other high support, and fasten the lower end to a hook in the floor so that the cord will be subjected to a very slight tension. Measure the length of the cord between the two supports. (The cord may be held to the floor by light pressure with the foot.)

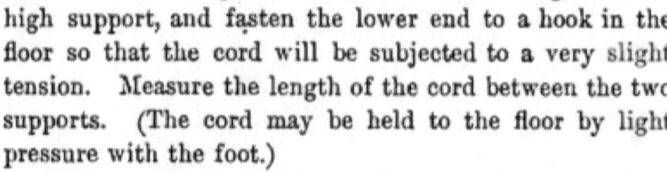

Fig. 21.

With a pointer or ruler tap the cord near the bottom; a short wave will run up the cord. Continue to tap the cord at regular intervals, timing the intervals so that the whole cord will vibrate as *one ventral segment*, as in Fig. 21.

When you have learned to keep the cord thus regularly in motion, count the number of *double* vibrations in 30 seconds. Repeat before recording.

Again, tap the cord at a much faster rate so it will vibrate in two ventral segments, with a *node*, *i. e.*, stationary point, at the middle of the cord. Count the number of vibrations now made in 30 seconds.

Similarly, make the cord vibrate in three, then four ventral segments, and determine the number of vibrations in 30 seconds in each case. Tabulate the results of the four tests as follows: —

	LENGTH OF SEGMENTS.	NUMBER OF VIBRATIONS IN 30 SECONDS.	NUMBER OF VIBRATIONS IN 1 SECOND.
1			
2			
3			
4			

What relation between length of segment and number of vibrations in a given time?

If on the water four waves, each 30 feet between crests, were to pass a given point in 1 minute, what would be the velocity per minute of the onward motion of the wave?

In your experiment each segment was but one-half the *wave length*. Knowing this, find the velocity of the wave motion per second in each of the above cases. Increase the tension and notice any effect this has on the rate.

EXERCISE 97. — Cause a bell to sound by striking it with a pencil or other light body; and while it is sounding suspend a small pith-ball or piece of cork from a thread so it rests lightly against the edge of the bell. What results? What seems to be the condition of the bell while sounding?

Tap a large tuning-fork on a piece of wood, and while sounding, invert it and dip the prongs about 1 mm. into water. What results? Sound again, and suspend the pith-ball against the prongs.

Stretch a short piece of No. 28 brass spring wire, or violin string, so that when plucked it will yield a musical tone, and while it is sounding, suspend a card or bit of paper lightly against the wire. What results? Why? Tear a strip of paper 1 cm. wide and 4 or 5 cm. long; then fold it so as to make a V-shaped saddle, technically known as a "rider." Now drop your rider astride the sounding string. What results?

Fasten the center of a 6-in. square pane of glass securely between two corks by means of a clamp. The edges of the glass should be ground or filed smooth. A brass plate, if not too thin, will answer still better. Sprinkle sand evenly over the surface of the plate, being careful that the plate is horizontal, then draw a violin bow across the middle of one edge of the plate so as to make a musical tone. What results? What must be the condition of the plate? Make a sketch of the plate after sounding a few times.

EXERCISE 98. — Hold a vibrating tuning-fork, whose pitch is not lower than middle C, over the mouth of an empty hydrometer jar about 15 or 16 in. deep. Does the sound appear louder than when the fork is held away from the jar? Again, bring the fork close to the mouth of the jar, and pour water slowly into the jar, listening carefully for the greatest intensity of sound. Try several times till you feel certain you have the point of greatest reënforcement; then measure the length of the column of air from the surface of the water to the mouth of the jar.

With the air column of the same length, rotate the vibrating fork slowly on its axis above the jar, and describe the effect. Now, starting from the position with one prong above the other, rotate through one-eighth of a turn, and state the effect; then through another eighth, etc., till a half-turn is completed. Call

these positions 1, 2, 3, 4 and 5, and designate the loudness of each. This change is due to *Interference* of the sound waves started by the two prongs.

The length of the air column giving greatest reënforcement, plus .3 the interior diameter of the jar, is one-fourth of the wave length of your fork. Knowing this, and the number of vibrations made by your fork in a second, determine the velocity of sound per second.

It will be well to repeat the experiment with a fork of different pitch, and compare with previous velocity.

EXERCISE 99. — To the bob of a heavy seconds pendulum a white cloth is tied so its vibrations can be seen at a distance. The pendulum is made to swing through a large arc, and just as it passes the middle of its arc, a person strikes a sharp rap on a board with a mallet or hammer, the motion of the hammer also being made visible by a white cloth. A second person seeks such a distance from the pendulum that the sound is heard just as the pendulum crosses the middle of the arc on its return swing. An opera-glass or spy-glass will be of great assistance. Begin with a distance of 850 or 900 ft., and increase till the proper point is found. As this method is at best not very accurate, great care should be exercised in timing the blows of the hammer to the swing of the pendulum. What do you find to be the velocity of sound? If there is a wind, exchange the positions of observer and pendulum, and find to what extent the wind increased or decreased the velocity of sound.

EXERCISE 100. — If a whirling table, with Savart wheel is at hand, hold the edge of a card against the teeth of the wheel, or hold the corner of the card so it will rub on the row of holes in the wheel while the wheel is in motion. A tone, more or less shrill, should be produced. Vary the rate of the wheel, and decide how the velocity with which the card is made to vibrate affects the pitch. If no whirling table is at hand, the escapement may be removed from an old clock, and a card held against the rapidly revolving wheel.

Hold a tube immediately over the outer row of holes of a Savart wheel, and blow steadily through the tube so that when the wheel is revolved, the air in passing from the tube through the holes will be broken up into puffs, and produce a musical tone.

How does the speed of revolution, and the consequent number of puffs per second, affect the pitch? Try over another row of holes.
Result?

If you have a good Savart wheel, it will be interesting to revolve till you bring the pitch in unison with a tuning fork; then, turning regularly at this rate for, say, ten seconds, determine the number of revolutions made by the crank on the wheel per second, then the number of revolutions made by the Savart wheel per second, and lastly, by counting the holes in the row used, the number of puffs per second which produced the tone.

EXERCISE 101. — *Apparatus:* A glass tube about 1.5 m. long and having an internal diameter of about 4 cm.; a brass tube or rod about 2 m. long and 1 cm. in diameter.

Fit to one end of the glass tube a stopper. Scatter through the tube a small quantity of cork dust. Fasten to the end of the brass tube, by means of sealing-wax, a thin cork that slides closely in the glass tube, but moves with slight friction. Push this into the glass tube about 50 cm. Lay all upon a table, and clamp the brass tube at its middle point by means of two grooved blocks and an iron clamp so that it will be held in a horizontal position and in line with the axis of the glass tube.

Rub a piece of resined leather over the outer end of the brass tube. A shrill sound should be produced and the cork dust be violently agitated. By moving the glass tube forward or backward so as to change the length of the confined air column, a point may be found where the sound will cause the cork dust to divide into segments separated by nodes. Find the average length of these segments.

Now these segments show the way in which the air in the closed tube is vibrating, and as the distance between two nodes in a vibrating body is half a wave length, we may thus find the wave length in air of the sound made by the brass. But the brass tube having a node in the middle is also half a wave length. By comparing the wave length in the brass tube with that in air, determine the relative velocities of sound in the two.

EXERCISE 102. — Apparatus: Sonometer with brass spring wire Nos. 22 and 28, B. & S. gauge.

Tune two No. 22 wires in unison so they will give a good musical tone; probably a tension of 15 to 20 lbs. will be needed. In

sounding a wire, pluck it in the middle, and, with the ear close to the wire, listen for the *fundamental* tone, which may for an instant be obscured by harsh or grating overtones.

Having the wires in unison, place a bridge under one a few cm. from the end, pluck the longer portion, and note any change of pitch. (It will be best to press the wire lightly against the bridge so as to limit the vibrations to the part under consideration.)

Now move the bridge till you find a length of wire whose tone is an octave above that of the full lengthed wire. Measure now the length of the wires.

Inasmuch as the higher tone of an octave is produced by twice as many vibrations per second as the lower tone, find from your results the relation between the number of vibrations per second and the length of the wire.

With same wires, apply tension of 6 lbs. to one, and if it does not produce a musical tone, place the bridge under *both* wires at such a point as will yield a good tone in the first wire. Now place a tension of 24 lbs. on the second wire and determine the musical interval between the tones produced by the two wires. What relation do you find between ratios of the numbers of vibrations per second and the ratios of the tensions?

Substitute a No. 28 wire for one of the No. 22 wires. Place a tension of 8 lbs. on each wire, and, if necessary, shorten *both wires equally* by means of the bridge so as to get a good tone from the thicker wire. Compare the vibration rates of the two wires. Compare the diameter of the wires (accurately measured, or taken from the tables) with the vibration rates.

Inasmuch as the wires are of equal length their weights will vary as their cross-sections, or as the squares of their diameters. Now find the relation that exists between the vibration rates and weights of the wires.

EXERCISE 103. — Tune two No. 22 wires in unison with tension of 15 to 20 lbs. While one is vibrating its full length, touch it lightly with a feather exactly in the middle; listen for the pitch of the resulting tone. Repeat, and drop a little paper rider upon the middle of the wire just after touching with the feather. What results? Try several times. What do you call the stationary points?

Now touch the wire while sounding its full length at a point one-third its length from the end. What tone results? Repeat, and

place two riders, one at the point touched, and one at the same distance from the other end. What results? Try several times. How does the wire seem to vibrate after touching? These higher tones are called *overtones*, or *harmonics*. What harmonic results when the wire is touched at one-fourth its length?

Now, while the wire sounds its fundamental, see if you can distinguish any of the harmonics without using the feather. Sound one of the wires vigorously, then stop it with the hand, and listen to see if any sound issues from the other wire. If so, what tone is it? The wires must be tuned exactly in unison to get results.

Two mounted tuning forks of same pitch will answer also. Sound one; stop it, and listen for the other. Vibrations thus produced are called sympathetic vibrations. Can you account for the way in which they are started? Now change the tension of one of the wires, or, if using the forks, fasten a drop of sealing-wax to the prong of one, and see if you can get the sympathetic vibrations. Explain.

Now, while the wires or forks are slightly out of tune, sound the two wires, or the two forks, and listen for beats, *i. e.*, swelling and diminishing of the tone. Count the number of beats in a given time, say 10 seconds. Then change the tension still more, or increase the weight of wax on the prong, and count beats again. How can you make use of beats in tuning an instrument?

EXERCISE 104. — Use tuning fork and two Y tubes connected as in the figure. A is a rubber tube 10 in. long; B is a bent glass tube 12 in. long capable of sliding inside the larger rubber tube C, which is 13 in. long; D need not be more than a few inches, but E should be 3 or 4 feet long.

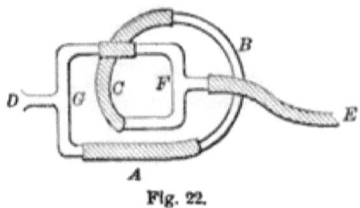
Fig. 22.

Let one person hold the free end of the tube E in his ear while another holds the vibrating fork at the open end of D. At first A, B, C should be drawn out full length. Now, seek to adjust the length of A, B, C, so that no sound is heard when the fork is held at D. Pinch the tube A shut; does the sound return? Try repeatedly. Having found a length of A, B, C, such that when A is closed, the fork is heard, but silence results on opening A, find the difference in the lengths of the two paths from F to G.

Compare this with the wave length of this fork as determined in Ex. 99. Try to account for the silence when both tubes are open.

The experiment may be varied by connecting both ends of a rubber tube 1.5 to 2 m. long to the arms of a Y tube; a second rubber tube connects with the other arm of the Y tube. While the free end of this last tube is held to the ear, the base of the vibrating fork is touched to the wall of the first tube at such a point as will produce silence. Find the difference in length of the two parts of this tube and compare this difference with the wave length of the fork as directed in the first part of this experiment.

LIGHT.

EXERCISE 105. — *Shadows.* Mount two cardboard screens about 3 cm. square and 20 cm. square. This may be easily done by fastening each with sealing-wax on a heavy wire and sticking this into a square block. Interpose the small screen between a flat gas jet and the larger screen, and adjust so the entire shadow of the first falls on the second. The flat flame may be conveniently obtained by unscrewing the tube from a Bunsen burner and screwing in its place a piece of gas tubing tipped with a regular gas burner.

Describe or sketch the shadow. Now, make a slight hole in the lighter part of the shadow called the penumbra. Look through it at the flame. If you see the flame, state what part of it. Now, make holes in the darkest part of the shadow, called the umbra; next in the line separating the umbra and penumbra; other places in the penumbra. State what you see in each case. From your observations, state the cause of the umbra and the penumbra.

Next turn the edge of the flame toward the small screen. Sketch the shadow. What sort of shadow would be made if the light came from a single luminous point? Test your last conclusion by turning the light very low. What sort of shadow have you? What difference have you noticed in the shadows made by an arc light and by a gas light? Account for the difference.

EXERCISE 106. — Mount three cardboard screens, 3 cm. (A), 6 cm. (B), and 9 cm. (C) square, with their centers the same height. Turn a gas jet very low and set it about 20 cm. from A, the center of the flame being the same height as that of the screen. Adjust B until the shadow of A just covers it. If A should now be removed, it is evident all the light that fell on it would then fall on B; as this is four times the size of A, the light on B would be only one-fourth as bright (intense) as on A. Or, representing the brightness (intensity) of the light on A by 1, the intensity on B will be .25.

Measure as carefully as possible the distance of the center of each screen from the center of the flame.

Now, take a new distance for A and adjust B as before. Try A with C; B with C; record your results as below, always representing the intensity of light on the screen nearer the flame by 1:

Screens Used.	Intensities.		Distances.	
	m.	n.	d.	D.
A and B				
A and B				
A and C				
A and C				
B and C				
B and C				

Comparing the results in the last two columns, state a law connecting the variation of intensity with variation of distance.

Exercise 107.—Make a Bunsen photometer as follows: Cut out of heavy unsized paper a circle about 10 cm. in diameter; drop in the center of it a little hot paraffine. Warm the paper gently until the paraffine has saturated the paper, making a translucent spot about 3 cm. in diameter. Mount this as the screens were mounted in the last Exercise. Do not, however, get any wax on the paraffined spot, or allow the wire to cross it. When this paper is illuminated more brightly on one side than on the other, the opaque ring appears brighter than the translucent center on the side of the brighter light, while on the darker side the reverse is true. When the two sides are equally illuminated, the spot has almost the same appearance as the rest of the paper. Mount on a block a candle, on another block four candles in a row close together. Set the single candle about 50 cm. from the paper so that the light from the flame shall fall perpendicularly on the center of the paper. Similarly on the opposite side set the row of four candles parallel to the paper, and move them until the paper, as tested by the translucent spot, appears exactly the same when viewed at the same angle on each side. Record distance. Make several trials and take the average. Since the two sides of the paper are now equally illuminated, how must the light which it receives from each candle of the four compare with that it receives from the single candle? How do their distances compare? What law does this approximately verify?

Compare the intensity of the light of a candle with that of your gas flame. Adjust them on opposite sides of your photometer until

they illuminate it equally. Measure their respective distances, and from these calculate the ratio of the intensity of the candle light to that of the gas light.

EXERCISE 108. — Mount two cardboard screens about 15 cm. square. Place these about 20 cm. apart and parallel to each other. Cut in the center of one a hole 2 mm. in diameter. Place 20 cm. from this a broad gas flame. Describe the image on the second screen. How does it compare in size with the flame? Can you account for its being inverted? Move either the flame or the screen. What change in the image? Why does not the flame imprint its image on all the surrounding surfaces? You may answer this by putting in the place of the screen with one hole another having many holes.

EXERCISE 109. — *Reflection from a Plane Mirror:* Through the middle of a large sheet of paper, say 50 cm. square, draw a line. On this line and parallel with it stand in a vertical plane a piece of looking-glass, about 10 cm. by 20 cm. If the glass is thick, it is well to recollect that the coated surface is the reflector. Stick a pin upright in the paper in front of the mirror. Sight along the edge of a ruler laid flat on the paper at the reflection of the pin. Draw a line AB along this edge; produce this line to intersect the line of the reflecting surface. Draw a second line, BC, from this intersection to the pin, and also from the same intersection draw a perpendicular to the line of the reflecting surface. The angle between the line from the pin and perpendicular is called the angle of incidence; the other is the angle of reflection. How do they compare in size?

Replacing the mirror, sight along the ruler from some other point and draw a line, EF, as before. Produce EF and AB until they meet. This point is the apparent position of the image of the pin. How do the distances of the pin and its image from the reflecting surface compare?

In front of the mirror draw a diagram, as a triangle, the outline of a face, etc. Place the pin upright at each of the angles of your diagram in turn and locate the position of the image of each point. Connect the points as located. How does the image so formed compare in size with the diagram? Since the image in a plane mirror has no real existence, it is called virtual or imaginary.

EXERCISE 110. — Hold a concave mirror in the sunlight so that the reflection from it is circular. Move a piece of chalk back and forth in front of the mirror until the point is found where this circle is smallest. This point is called the principal focus of the mirror. Its distance from the mirror is called the focal length of the mirror; measure it carefully and record.

Second Method: — Stand alongside of a flat gas jet a cardboard screen, both being in the same plane. Move the mirror back and forth in front of these until you obtain a sharp image of the flame on the screen. The screen, or flame is now very nearly at the center of the sphere of which the mirror is a part. What is the radius of this sphere? Half this is the focal length of the mirror. Record it.

(a) Is the image formed on the screen virtual or real?
(b) Erect or inverted?
(c) How does it compare in size with the object?

Place the flame between the center of curvature (center of the sphere) and principal focus of the mirror, adjust a screen until a sharp image is formed. Note distances of object (flame) and image from the mirror. Describe the image according to (a), (b) and (c) above. Without moving the mirror, let flame and screen change places. Describe image by (a), (b) and (c).

Points near a lens or mirror related to each other, as the positions of the flame and image in this experiment, are called conjugate foci. Formulate a definition for them.

Move the flame nearer the mirror than the principal focus. Can you get any image on a screen? Describe, as before, the image in the mirror.

Place the flame at any point in front of a convex mirror, and describe the image obtained.

EXERCISE 111. — Place a bright coin on the bottom of a pan. Stand in such a position that the coin is just hidden by the side of the pan. Without moving, watch while another pupil pours water into the pan. Note the depth of the water when the coin appears to you, if it does so. Remembering that we see objects by the rays of light passing from them to our eyes, and that the rays from the coin could not reach the eye before the water was poured into the pan, draw a diagram showing what effect the water must have had on the path of some of the rays from the coin.

EXERCISE 112. — Hold a bi-convex lens in the sunlight so that a circle of light is cast on a screen behind it. Move the screen until the circle is the smallest possible; this is the point to which the lens by refraction converges the parallel rays of the sun, and is called the principal focus of the lens, and the distance of this point from the nearer face of the lens is the focal length. Record it. Obtain on a screen the image of some distant object. How does the distance of this image from the lens compare with the focal length previously found?

Second Method: — Adjust a bright gas flame and a screen on opposite sides of a lens so that the image of the flame is cast on the screen, and both are at the same distance from the lens. Both are now at points called secondary foci of the lens. The principal focus is midway between the secondary focus and the lens. Record the focal length of your lens. Compare the image and object in this Exercise according to (*a*), (*b*) and (*c*) of Exercise 110.

Place the flame between the principal and secondary foci and move the screen until a sharp image is obtained. Note distance of screen and object from the lens, and describe the image as before.

Let screen and flame exchange places. Describe image as before.

What name may be given to the points where the flame and image stand?

Place the flame nearer the lens than the principal focus. Can you get any image on the screen? If, by looking through the lens at the flame, an image of the flame is seen, describe it.

Place the flame anywhere before a concave lens and describe the image.

EXERCISE 113. — Mount two small bi-convex lenses about 3 cm. and 5 cm. focus, and 4 cm. diameter, by sticking the edge of each into a slit in a large cork. Hold some small bright object in front of and near to the lens of shorter focus. Adjust behind it a screen so as to obtain a sharp image of the object. Beyond this screen and nearer to it than the focal distance of the second lens place the second lens. What sort of an image will a convex lens produce of anything nearer to it than its principal focus? Now remove the screen, and look through the second lens toward the first. You will now likely see a magnified virtual image of the true image formed by the first lens; a little more adjustment of

the lenses may be necessary to make this image distinct. This is the principle of the compound microscope. The lens nearer the object is the objective; the other, the eye-piece.

Adjust as in the previous experiment two lenses, using for the objective a lens of 40 cm. focus and 10 cm. diameter, and for the eye piece the lens of 3 cm. focus and 4 cm. diameter. Obtain on your screen, however, the image of some distant object; then place your eye piece in the proper position and describe the result. This shows the action of a compound telescope.

American Wire Gauge. (B. & S.)

No.	Diam. in mm.	Resistance in Ohms per 1000 ft.	Resistance in Ohms per 1000 m.	No.	Diam. in mm.	Resistance in Ohms per 1000 ft.	Resistance in Ohms per 1000 m.
1	7.348	.129	.423	21	.723	13.323	43.699
2	6.544	.163	.534	22	.644	16.799	55.090
3	5.827	.205	.672	23	.573	21.185	69.486
4	5.189	.259	.849	24	.511	26.713	87.618
5	4.621	.326	1.059	25	.455	33.684	110.483
6	4.115	.411	1.348	26	.405	42.477	141.324
7	3.665	.519	1.702	27	.361	53.563	175.886
8	3.265	.654	2.145	28	.321	67.542	221.537
9	2.907	.824	2.702	29	.286	85.170	279.357
10	2.588	1.040	3.411	30	.255	107.391	352.242
11	2.305	1.311	4.300	31	.227	135.402	444.118
12	2.053	1.653	5.421	32	.202	170.765	540.109
13	1.828	2.084	6.835	33	.180	215.312	706.223
14	1.628	2.628	8.619	34	.160	271.583	890.822
15	1.450	3.314	10.962	35	.143	342.443	1133.213
16	1.291	4.179	13.706	36	.127	431.712	1416.015
17	1.150	5.269	17.282	37	.113	544.287	1785.261
18	1.024	6.645	21.780	38	.101	686.511	2250.756
19	.899	8.617	28.265	39	.090	865.046	2837.350
20	.812	10.566	34.656	40	.080	1091.865	3581.317

The resistances given are for *pure copper* wire at a temperature of 24° C. (75° F.). Ordinary copper wire has a resistance 5 or 6 per cent. higher than pure copper.

Formulae for Calculating Areas and Volumes.

Area of a circle $= \pi R^2$, $\pi = 3.1416$, $R =$ radius.
Volume of a cylinder $=$ area of the base $\times$ the altitude.
Volume of a sphere $= \frac{1}{6} \pi D^3$, $D =$ diameter.

Table of Specific Gravities.

Alcohol, absolute	0.806	Kerosene	0.810
" common	0.833	Lead, sheet	11.400
Alum	1.724	Limestone	3.180
Ashwood	0.690	Marble	2.720
Beeswax	0.964	Mercury	13.596
Brass, cast	8.400	Milk	1.032
Coal, bituminous 1.270 to	1.423	Oak, red	0.850
" anthracite 1.260 to	1.800	" white	0.779
Cork	0.240	Oil, olive	0.915
Copper	8.850	" turpentine	0.870
Feldspar	2.600	Paraffin 0.820 to	0.940
Galena	7.580	Pine, white	0.554
German Silver	8.432	" pitch	0.060
Glass 2.50 to	3.600	Platinum Wire	21.530
Gold	19.360	Quartz	2.650
Granite	2.650	Silver, cast 10.400 to	10.510
Gutta-percha	0.966	Steel	7.816
Ice	0.917	Sulphur, native	2.030
Iron, cast	7.230	Sulphuric Acid	1.840
" wrought	7.780	Tin, cast	7.290
India-rubber	0.930	Vinegar	1.026
Iron Pyrites	5.000	Zinc, cast	7.000

Table of Trigonometrical Functions.

Deg.	Tang.	Sine.	Deg.	Tangent.	Sine.	Deg.	Tangent.	Sine.
1	.017	.017	31	.601	.515	61	1.80	.875
2	.035	.035	32	.625	.530	62	1.88	.883
3	.052	.052	33	.649	.545	63	1.96	.891
4	.070	.070	34	.675	.559	64	2.05	.899
5	.087	.087	35	.700	.574	65	2.14	.906
6	.105	.105	36	.727	.588	66	2.25	.914
7	.123	.122	37	.754	.602	67	2.36	.921
8	.141	.139	38	.781	.616	68	2.48	.927
9	.158	.156	39	.810	.629	69	2.61	.934
10	.176	.174	40	.839	.643	70	2.75	.940
11	.194	.191	41	.869	.656	71	2.90	.946
12	.213	.208	42	.900	.669	72	3.08	.951
13	.231	.225	43	.933	.682	73	3.27	.956
14	.249	.242	44	.966	.695	74	3.49	.961
15	.268	.259	45	1.000	.707	75	3.73	.966
16	.287	.276	46	1.036	.719	76	4.01	.970
17	.306	.292	47	1.070	.731	77	4.33	.974
18	.325	.309	48	1.110	.743	78	4.70	.978
19	.344	.326	49	1.150	.755	79	5.14	.982
20	.364	.342	50	1.190	.766	80	5.67	.985
21	.384	.358	51	1.230	.777	81	6.31	.988
22	.404	.374	52	1.280	.788	82	7.12	.990
23	.424	.390	53	1.330	.799	83	8.14	.993
24	.445	.407	54	1.380	.809	84	9.51	.995
25	.466	.423	55	1.430	.819	85	11.43	.996
26	.488	.438	56	1.480	.829	86	14.30	.998
27	.510	.454	57	1.540	.839	87	19.08	.999
28	.532	.469	58	1.600	.848	88	28.64	.999
29	.554	.485	59	1.660	.857	89	57.29	1.000
30	.577	.500	60	1.730	.866	90	Infinite.	1.000

INDEX.

EX.	MAGNETISM.	PAGE
1.	Magnetization	2
2.	Magnetic attraction without contact	4
3.	Locating the poles of a magnet	4
4.	Naming the poles of a magnet	6
5.	Construction and use of a magnetoscope	6
6.	Effect of magnetic poles on each other	6
7.	Effect of breaking a magnet	8
8.	Arrangement of poles in magnetization	10
9.	Reversal of polarity	10
10.	Induced magnets	10
11.	Magnetic fields	12

MEASURING.

12.	Measuring lengths and plane surfaces	14
13.	Measuring volumes	14
14.	Using micrometer calipers	14
15.	Weighing	14

PROPERTIES OF MATTER.

16.	Impenetrability	16
17.	Porosity	16
18.	Ductility	16
19.	Inertia	16
20.	Elasticity of air	18
21.	Elasticity of rubber; of stretching wires	18
22.	Testing the stiffness of wooden rods	22
23.	Elasticity of torsion of wooden rods	24
24.	Tenacity of wires	26
25, 26.	Cohesion and adhesion	28
27.	Capillary action	30

EX.	PENDULUMS.	PAGE
28.	Pendulums	34

MECHANICS.

29.	Parallel forces in the same plane	38
30, 31, 32.	Levers	40–44
33.	Wheel and axle	46
34.	Pulleys	46
35.	Inclined plane	48
36.	Friction	50
37.	Center of gravity	52
38.	Influence of the weight of the lever	52
39.	Resultant of angular forces	52
40.	Laws of falling bodies	54

HYDROSTATICS.

41.	Liquid pressure	58
42.	Archimedes' principle	58

SPECIFIC GRAVITY.

43.	Of a rectangular solid	62
44.	Of an irregular solid denser than water	62
45.	Of an irregular solid less dense than water	62
46.	Of a liquid by specific gravity bottle	62
47.	Of a liquid by balanced columns	64
48.	Of a liquid by simple hydrometer	64

PNEUMATICS.

49.	Specific gravity of air	66
50.	Determine the value of air pressure	66
51.	Lifting pump, force pump	66
52.	Siphon	68
53.	Mariotte's law	68

HEAT.

EX.		PAGE
54.	Heat from mechanical energy	70
55.	Expansion of air	70
56.	Expansion of water	70
57.	Coefficient of linear expansion	72
58.	Conduction of heat	74
59.	Conductivity of water	74
60.	Convection in water	76
61.	Convection in air	76
62.	Freezing-point of water	76
63.	Boiling-point of water	78
64.	Effect of pressure on boiling-point	78
65.	Boiling-points of ether and turpentine	78
66.	Cooling processes	80
67.	Dew point	80
68.	Law of cooling	82
69.	Latent heat of water	84
70.	Latent heat of steam	84
71.	Specific heat of shot	86

ELECTRICAL ENERGY FROM FRICTION.

72.	Development of two kinds of electrification	88
73.	The electroscope, its construction and use	88
74.	Charging by induction	90
75.	Conductors and insulators	90
76.	Potential	90
77.	Resistance	92
78.	Electro-motive force	94
79.	Condensers, Leyden jar	94

CURRENT ELECTRICAL ENERGY.

80.	A galvanic cell	98
81.	Names of plates and electrodes	98
82.	A two-fluid cell	100
83.	A gravity cell	102
84.	Effects of an electrical current	102
85.	Electro-magnets	104
86.	Ampere's law	106
87.	The essentials of a galvanometer	106

EX.		PAGE
88.	A tangent galvanometer	108
89.	Tests for conductivity of electrical currents	110
90.	Effects of a neighboring magnet on galvanometer deflection	112
91.	Measuring resistances by substitution	112
92.	Measuring resistance of a battery cell	114
93.	Resistance by Wheatstone's bridge	116
94.	Induced currents; idea of the dynamo	118
95.	Development of the idea of the electro-motor	120

SOUND.

96.	Vibrations in a rubber cord	124
97.	Vibrations of a sounding body	126
98.	Reinforcement, interference and velocity of sound	126
99.	Velocity of sound	128
100.	Determining pitch	128
101.	Velocity of sound in a brass rod	130
102.	Laws of vibrating strings	130
103.	Harmonics and beats	132
104.	Interference of sound	134

LIGHT.

105.	Shadows	138
106.	Law of intensity	138
107.	Photometry	140
108.	Images through apertures	142
109.	Reflection from a plane mirror	142
110.	Reflection from spherical mirrors	144
111.	Refraction by water	144
112.	Refraction by spherical lenses	146
113.	Principle of the compound microscope; principle of the compound telescope	146

www.ingramcontent.com/pod-product-compliance
Lightning Source LLC
Chambersburg PA
CBHW030309170426
43202CB00009B/924